Construction Adjudication in Ireland

The Construction Contracts Act 2013 introduces adjudication for the construction industry in Ireland for the first time. The essence of adjudication is in providing a means whereby disputes as to payment under a construction contract are resolved quickly and cheaply. The key feature distinguishing adjudication from other processes is that the money found due by the adjudicator must be paid pending the outcome of arbitration or litigation. Its primary function, therefore, is to ensure cash flow for contractors and sub-contractors.

Leading construction lawyer Anthony Hussey's new book is the first to provide a section by section analysis of the Act itself, an analysis of the Code of Practice, and a discussion of the likely constitutional issues to which the legislation will give rise.

This practical legal reference is aimed at all those involved in construction contract disputes, be they lawyers, architects, engineers, quantity surveyors, contractors or sub-contractors.

Anthony Hussey specialises in Construction Law and acts mainly for contractors and sub-contractors. His expertise in this regard is predominantly in the area of dispute resolution but he also advises on/drafts contract documents and issues of procurement law. He has previously lectured in the law of Contract and Tort for a postgraduate course at Trinity College, Dublin, Ireland, and was also external examiner to the postgraduate Construction Law course run by the engineering faculty of Trinity College.

Construction Adjudication in Ireland

Anthony Hussey

Routledge
Taylor & Francis Group

LONDON AND NEW YORK

First published 2017 by Routledge

2 Park Square, Milton Park, Abingdon, Oxfordshire OX14 4RN
52 Vanderbilt Avenue, New York, NY 10017

Routledge is an imprint of the Taylor & Francis Group, an informa business

First issued in paperback 2020

British Library Cataloguing-in-Publication Data
A catalogue record for this book is available from the British Library

Library of Congress Cataloging in Publication Data
Hussey, Anthony, author.
Construction adjudication in Ireland / Anthony Hussey.
pages cm
Includes bibliographical references and index.
1. Construction contracts—Ireland. 2. Arbitration and award—Ireland.
I. Title.
KDK660.H87 2016
347.41707'8624—dc23
2015036975

ISBN: 978-1-138-18792-4 (hbk)
ISBN: 978-0-367-59555-5 (pbk)

Typeset in Times
by Swales & Willis Ltd, Exeter, Devon, UK

Contents

4 Payment entitlements   **24**

5 Payment claim notices   **36**

6 The adjudication process   **50**

Preface

The text of this book has been changed on a number of occasions to allow for different drafts of the Code of Practice and ultimately the Statutory Instruments giving legal force to the Code published on the 5th July 2016 and revoked and republished on the 25th July 2016.

As of the 26th July 2016 the Act is in force and applies to all contracts entered into after the 25th July 2016. The Code of Practice is in force and new Rules of Court have been introduced. Six years after inception, and three years after being signed by the President, the legislation is finally ready to make its impact. On the whole it is being welcomed by the industry, albeit more warmly by sub-contractors than main contractors. The industry desperately needs a mechanism for resolving disputes other than arbitration or litigation. Disputes in the industry are inevitable and common place. The industry cannot afford to have arbitration or litigation as the final resort. Although adjudication is not final in theory, in practice it does provide the final solution in the vast majority of cases.

The legislation is in many respects flawed in its detail. The big challenge for the industry is to persuade the Courts to uphold and support the legislation notwithstanding these flaws and the inherent resistance to a system which openly provides for rough justice albeit, in theory at any rate, on a temporary basis.

Throughout the text I have used abbreviated descriptions for legislation and reports. These are listed in an appendix. I have also for brevity described the laws of England and Wales and those of Scotland in relation to adjudication in a generic manner given that there is, on this issue, very little difference between the two.

This book is intended for a wide audience comprising of developers, contractors, sub-contractors, engineers, architects, quantity surveyors, lawyers and others involved in construction projects. In so far as it necessarily involves issues of legal interpretation, it does rely to some extent on comparative analyses of the case law of other jurisdictions. I hope however that I have succeeded in applying a light touch in that regard.

I would like to thank Niall Lawless for the inspiration and encouragement. I would also like to thank my colleagues at Hussey Fraser, Simon Fraser and Siobhan Kenny for their research and contribution and Sandra Shanahan for her patience and exceptional competence in pulling it all together. Last, but by no means least, my thanks to my wife Ursula for her constant support and tolerance.

August 2016
Anthony Hussey

Foreword

The Construction Contracts Act 2013 introduces into the Irish legal landscape a new method of resolving certain payment disputes arising out of construction contracts. When new legislation is introduced, it is always helpful for both legal practitioners and parties working in the relevant field to have access to an informed commentary on the new enactment with helpful guidelines as to how it might operate by reference to other comparable jurisdictions where similar schemes are in place. The author of this text, Mr Anthony Hussey, is an experienced lawyer and arbitrator in the field of construction contracts. He brings his knowledge and expertise to bear on the subject by presenting the reader with a helpful overview of the new legislation and a comprehensive analysis and commentary on each section of the Act. In the absence, as yet, of any Irish jurisprudence on the Act, he makes good use of case law and other materials from jurisdictions with comparable legislation, so as to inform the reader as to the likely effects of the legislation in this state.

The importance of early payment of building contractors and sub-contractors and an efficient and cost-effective resolution of disputes surrounding payment cannot be overestimated. Sadly, it is not uncommon for efficient and capable contractors and sub-contractors to go out of business, because they have not been paid for their work in a timely manner. This has far-reaching consequences for not only the parties concerned and their employees but has a wider knock-on effect on the general economy, which depends on an efficient and productive construction industry. The Irish legislation draws a distinction between main contractors and sub-contractors in so far as measures for determining the timing and amount of payments are concerned. This is different to the legislation in the UK and other jurisdictions and is one of a number of important topics, which are addressed by the author in the text.

In this work, Mr Hussey deals with the scope and ambit of the Act in a comprehensive way. The text sets out in a clear and logical manner references to the Act, with informed commentary and helpful references to guide the reader through the new regime with confidence. From the scope of the Act to the enforcement by the courts of adjudicators' decisions, the new legal landscape is mapped out clearly and precisely. Mr Hussey has done a great service to the legal profession and to all involved in the construction industry in bringing this work to publication.

Brian J. McGovern
High Court
Four Courts
Dublin 7

July 2016

Abbreviations

The UK Act means The Housing Grants, Construction and Regeneration Act 1996;

The Amending Act of 2009 means the Local Democracy, Economic Development and Construction Act 2009;

The Scheme means The Scheme for Construction Contracts (England and Wales) Regulations 1998, The Scheme for Construction Contracts (England and Wales) 1998 (Amendment) (England) Regulations 2011, The Scheme for Construction Contracts (Scotland) Regulations 1998, the Scheme for Construction Contracts (Scotland) Amendment Regulations 2011;

The New South Wales Act or NSW Act means the Building and Construction Industry Security of Payment Act 1999 as amended most recently in 2014;

The Victorian Act means the Building and Construction Industry Security and Payment Act 2002;

The New Zealand Act means the Construction Contracts Act 2002;

The Isle of Man Act means the Construction Contracts Act 2004;

The Northern Territory Act means the Construction Contracts (Security of Payments) Act 2004;

The Singapore Act means the Building and Construction Industry Security of Payment Act 2004;

The Western Australian Act means the Construction Contracts Act 2004;

The Queensland Act means the Building and Construction Industry Payments Act 2004 as amended by legislation up to 2013;

The Tasmanian Act means the Building and Construction Industry Security of Payments Act 2009;

The Australian Capital Territory Act means the Building and Construction Industry Security of Payment Act 2009;

The Malaysian Act means the Construction Industry Payment and Adjudication Act 2012;

The Irish Act means the Construction Contracts Act 2013.

Reports

The Wallace Report means Discussion Paper – Payment Dispute Resolution in the Queensland Building & Construction Industry, Final Report, Andrew Wallace, Barrister at Law, May 2013.

The Collins Report means The Final Report of the Enquiry into Construction Industry Insolvency in New South Wales, Chaired by Mr Bruce Collins Q.C., January 2013.

Table of cases

1   Introduction

Following the collapse of the Irish banking system and consequently the Irish economy in 2008, a large number of developers and main contractors in the construction industry became insolvent. The insolvency of the main contractors highlighted the extent of their indebtedness to numerous sub-contractors. It was clear that a culture of payment deferral in respect of sub-contractors had been in operation for many years, with the effect that the insolvency of the main contractors not only meant that sub-contractors were not paid in relation to recent developments but were also not paid for work carried out in earlier days when the industry was at its most profitable. Sub-contractors pressed hard for the introduction of legislation which would ensure that sub-contractors would be paid promptly as a matter of right. In the years following the collapse of the economy, the Government, though approving of the concept, was not prepared to devote the time necessary for the introduction of legislation. Seanad Eireann, the upper house of parliament, does have a rarely used entitlement to introduce legislation. A senator, Fergal Quinn, took a special interest in the issue and in May 2010 introduced through the Senate the Construction Contracts Bill. Once introduced, the Government, and indeed also the opposition, pledged to support the Bill. Notwithstanding that support, the legislation in the form of the Construction Contracts Act (hereinafter referred to as the **Irish Act**) was not passed into law until 2013. Although passed into law, its commencement was subject to an appropriate ministerial order being made. This order was made in April 2016; it provided that the Act would apply from the 25th day of July 2016.

The Irish Act essentially does two things. First, it provides a mechanism whereby main contractors may, and sub-contractors will, be paid promptly for the value of their work as the contract works proceed. Second, it introduces adjudication in relation to payment disputes with a view to ensuring that payment cannot be unduly delayed simply because it is disputed in whole or in part.

Adjudication is arguably the most radical interference by the legislature of any country in which it has been introduced in respect of the right to contract on such terms as the parties deem appropriate, with the possible exception of legislation prohibiting unfair terms where such legislation applies. The legislation providing for adjudication was first introduced in the UK in 1996, New South Wales followed in 1999, and a number of other Australian states subsequently passed similar legislation. Adjudication legislation is also in force in New Zealand, the Isle of Man, Singapore and Malaysia. In all of these jurisdictions, adjudication is an option and is not a mandatory process for the resolution of disputes. It is availed of extensively, particularly in the UK and in Australia.

The purpose of adjudication in the words of Dyson J in *Macob Civil Engineering Limited v Morrisson Construction Limited*:[1] 'Was to introduce a speedy mechanism for settling disputes in construction contracts on a provisional interim basis, and requiring the decisions of adjudicators to be enforced pending the final determination of disputes by arbitration, litigation or agreement'.

A different, but equally valid, description of adjudication was given by Lord Ackner at the report stage in respect of the legislation in the House of Lords: 'Adjudication is a highly satisfactory process. It comes under the rubric of "pay now, argue later" which is a sensible way of dealing expeditiously and relatively inexpensively with disputes which might hold up important contracts'.[2]

In the countries in which it has been introduced, adjudication is not an option under the legislation in respect of every construction contract. The legislation everywhere provides for exceptions. However, once the concept of adjudication is introduced by legislation, there has been a tendency to amend standard forms of contract to provide for adjudication. One of the contracts excluded by the UK Act of 1996 relates to residential dwellings. It was held by the Technology and Construction Court in *Picardi v Cuniberti*[3] that a contractual requirement for adjudication in respect of such an excluded project was not an unfair term and accordingly was enforceable. Therefore, adjudication, while only an option under the legislation, may very well be compulsory by virtue of the terms of contract between the parties. If adjudication takes place pursuant to the terms of the contract, it is not statutory adjudication, but subject to the procedural rules incorporated in the terms of contract, which in turn are usually very closely aligned to statutory adjudication.

The Irish Act is a very simple one comprising 12 sections and 13 pages. The fact that it took three years to pass the legislation when there was no opposition to it, and that even then it took a further three years to put in place the measures necessary for its implementation, suggests that the legislative system is far from perfect.

Unfortunately, the Irish Act itself is also far from perfect. This was recognised by Senator Quinn who instigated the legislation and by the Construction Industry Federation who lobbied to have the legislation dealt with effectively and efficiently. When the Government first took the Bill under its wing, it sought to introduce safeguards it thought to be appropriate that would, if introduced, have wholly defeated the purpose of adjudication. These included a provision that adjudication would only apply in respect of contracts with a value in excess of €200,000, and that an adjudicator's determination would only be binding if a bond was provided for the sum of money to be paid under it. It took so much time and effort to persuade the Government to withdraw these provisions that the promoters of the legislation were inclined to live with any other defects rather than delay the matter further.

Section 9 of the Irish Act provides that the relevant minister: 'may prepare and publish a Code of Practice governing the conduct of adjudications'. In reality, the legislation is incomplete without such a Code of Practice being published. A number of draft of the Code were circulated for comment by the Department to the relevant stakeholders prior to the Code of Practice being finally published the 25th July 2016.

Under section 8 the Minister was to appoint adjudicators from members of a panel to be created by the Minister. It was necessary to put in place the panel of adjudicators and to appoint a chairperson in relation to it prior to the implementation of the legislation. The panel of adjudicators has been completed. The list of panel members is available on the Iris Oifigiuil Website (http://www.irisoifigiuil. ie/archive/2016/january/IR150116-2.pdf).

The elephant in the room, in so far as the Irish Act is concerned, is the Irish Constitution of 1937. It is conceivable that the Irish Act would be struck down in its entirety on the basis that the Irish Act does not incorporate sufficient (or any) safeguards to ensure that the requirements of constitutional justice will be met. This is considered unlikely. However, it is likely that the Irish courts on constitutional grounds will balk at the prospect of enforcing adjudicators' decisions that are clearly incorrect. The courts in the UK and elsewhere have robustly supported adjudication to the extent of enforcing decisions that were manifestly incorrect. The difficulty is that once one departs from this principle it is challenging to set a limit as to the extent to which the courts should go in ascertaining whether or not the decision is incorrect. The test under the Arbitration Acts 1954–1998 was whether there was a fundamental error on the face of the award. Ultimately, however, the courts arguably went beyond that test and significantly so in the Supreme Court judgment in *Galway City Council v Samuel Kingston Construction Ltd*.[4]

The Code of Practice requires that adjudicators' decisions be reasoned. There will always therefore be a basis for arguing that the adjudicator's decision is wrong. If the courts are prepared to set aside a decision on that basis alone, the whole purpose of the exercise could be defeated, i.e. a temporary entitlement to payment both quickly and inexpensively.

The real difficulty for the Irish courts may centre on the constitutional right to have issues of fact in dispute tested through oral evidence and cross examination. If that entitlement is assiduously applied, it may be possible for a respondent to render the process incapable of practical application.

Whereas the wording of the Irish legislation is not above criticism, in the bigger picture this is likely to be of little consequence. The expectation is that the very implementation of the Act will change the culture and, in particular, the treatment by main contractors of sub-contractors. This has been the experience elsewhere. One commentator's appraisal of the New Zealand Act some four years after its implementation was summarised as follows:

> The Act's biggest effect has, undoubtedly, been simply because it exists. Employers and contractors (at least those who are familiar with the Act) are, I believe, being more sensible about payment than they were or may have been previously, because they do not wish to get involved in an adjudication.[5]

The fact is that elsewhere, adjudication has changed the culture partly because disputes referred to adjudication rarely go beyond that point to arbitration or lit-iga-tion. This was recognised as early as 2002 by his Honour Judge Humphrey Lloyd:

> It is now clear that the construction industry regards adjudication not simply as a staging post towards the final resolution of the dispute in arbitration or litigation but as having in itself considerable weight and impact that in prac-tice goes beyond the legal requirement that the decision has for the time being to be observed.[6]

Throughout this book reference is made to the legal position in the UK as if there was only one law applicable in the UK. This is because the law in relation to adju-dication is to all intents and purposes the same in all parts of the United Kingdom. The legislation providing for adjudication, not just in the UK, but elsewhere, tends to be titled in a particularly wordy fashion. When referring to legislation, the author has preferred to refer to the country or state of origin, such as the New South Wales Act rather than its official title, i.e. Building & Construction Industry Security of Payment Act 1999. A number of abbreviated descriptions have been used for this and other purposes. A full list of these is included at page xiii.

International context

The purpose of legislation introducing adjudication in most countries was similar to that which motivated the Irish Government – a desire to ensure fluidity in cash flow and, in particular, to ensure that sub-contractors would have the mechanism available to secure early payment for their work. In summing up the requirement for such legislation across Australia, one commentator put it as follows:

> Cash flow is especially pertinent for smaller sub-contractors who rely on it to meet debt obligations and keep their businesses solvent. Made vulnerable by their dependence on payment, these sub-contractors can be taken advan-tage of by upstream debtors seeking to increase their margins by deliberately withholding payment in the hope that their creditor will become bankrupt. Recognising that these practices could not be allowed to prevail, govern-ments took action to address the problem.[7]

This legislation, across the world, tends to include provisions which are not strictly related to adjudication. For instance, nearly all contain provisions prohibiting *pay when paid* type clauses. Some also seek to secure payments due to contractors and sub-contractors through trust provisions, and some also seek to empower a sub-contractor, of whatever tier, to obtain payment directly from the party who employs the party with whom it has a contract. This publication, however, is not concerned with these peripheral issues. It is concerned with adjudication and the payment provisions linked to adjudication through the Irish legislation.

On the whole, the legislation in the European countries (the UK, the Isle of Man and Ireland) tend to be similar in terms of the underlying principles and noticeably different to the legislation of the non-European countries, i.e. Australia, Singapore, Malaysia and New Zealand. The following comparisons arise:

- The legislation provides a definition for the construction contracts to which it applies. This is common to all countries and there are subtle differences in each jurisdiction.
- The term *construction contracts* is defined in most of the non-European countries to include the supply of materials and components for installation through a construction contract. Supply contracts are excluded by the European countries.
- The legislation provides for payment claims and responses to payment claims. This would be a common feature in both the European and the non-European countries.
- If the respondent fails to respond to the payment claim within the time allowed, either the sum due under the contract or the amount claimed becomes due. Until the UK Act of 2009, the position in the European countries would have been uniform, i.e. it was only the amount due under the contract that could be recovered by the claimant. The legislation of the non-European countries generally provides that it is the amount claimed that becomes due.
- The legislation of the non-European countries provides that if the amount claimed, or the amount not in dispute if the respondent delivers a response, is not paid, summary judgment may be obtained. This is absent from the legislation of the European countries.
- Disputes may be referred to adjudication. Some countries provide that only payment disputes may be so referred and others provide that any dispute may be referred to adjudication. There is no distinct divide between Europe and elsewhere in this regard.
- In the European countries, a party is entitled to seek adjudication at any time. In the non-European countries a party may lose the entitlement if it does not seek adjudication within a specific time after the work has been completed.
- The non-European countries provide for the respondent making a response to the claimant's adjudication notice (sometimes called adjudication application or referral). The European countries do not provide for this – how the procedure is to be run after the claimant's referral has been made is left entirely to the adjudicator.
- In the non-European countries the time the adjudicator has to make a decision runs from the date that the respondent is required to deliver its response document. In the European countries it runs from the date the referral is delivered to the adjudicator.
- The adjudicator makes a decision within the time required and that decision is binding pending the dispute being resolved through litigation or arbitration as appropriate. This is common to all.

- The legislation of the non-European countries tends to state that adjudicators are not entitled to their fees if they fail to deliver their decision on time. This is not provided for in the legislation of the European countries.
- The non-European countries provide that adjudicators will not be in breach of their obligations if they retain the decision pending payment of their fees. This is not provided for in the European countries.

The first legislation to be passed in Australia after the UK Act of 1996 was the Building and Construction Industry Security and Payment Act 1999 of New South Wales. Since then, similar legislation has been introduced in Victoria, Queensland, Northern Territory, Western Australia, Tasmania, Australian Capital Territory and South Australia.

The construction industry payment legislation across Australia has been broadly categorised into two camps, being the East Coast and the West Coast models.[8] The West Coast for this purpose comprises the Northern Territories and Western Australia, the remaining states and territories making up the East Coast model. The West Coast model was considered to be more in harmony with the legislation passed in the UK and New Zealand and can be distinguished from other Australian legislation, which more closely resemble the NSW Act (the East Coast model). One of the main differences between the two models is that the East Coast model is more inclined to allow the parties to decide through their contract when payments would be made, whereas the West Coast model provides for maximum intervals between progress payments and for the payment of the final account, which would be mandatory unless the contract provided for lesser periods. Furthermore, the East Coast model only allows for adjudication up the contractual stream, that is to say a sub-contractor could pursue a main contractor and a main contractor could pursue an employer through adjudication, but not vice versa.

Considerable thought has been put into recent amending legislation to the East Coast States of New South Wales and Queensland. The New South Wales Amendment Act of 2013 followed a detailed report prepared by Bruce Collins QC,[9] and the Queensland Building & Construction Industry Payments Amendment Bill of 2014 followed after a detailed review of the operation of adjudication in that state by Andrew Wallace.[10] Aspects of these reports will be discussed in the context of the Irish legislation as appropriate, but it is worth noting at this stage a unique aspect to the Queensland reforming legislation. On the recommendation of the Wallace Report, it introduces different timeframes for dealing with complex claims and standard claims. When applying for adjudication, the claimant must specify whether the claim is a standard claim or a complex claim. The latter is a claim involving a sum in excess of $750,000, a claim in relation to latent defects or a time-related claim.

Unfortunately, these reports would not have been available to the draftsman at the time that the Irish legislation in its final form was being considered. It would appear that the draftsman of the Irish legislation primarily relied upon the UK experience.

It is somewhat of an oddity that a country such as Australia, with a population of about 23 million people, has 8 separate enactments applying across the land, whereas hugely more populated regions such as the UK and Malaysia have, essentially, only one statute each. The problem is exacerbated in Australia by the fact that the legislation in each state or territory is different and, in subtle respects, has been interpreted differently by the courts giving rise to confusion wherever inter-state development occurs. There have been many papers written exhorting harmonisation of the legislation across Australia.[11] The fact, however, that there are eight different legislative enactments, each with its own idiosyncrasies, and based on two entirely different philosophies, gives rise to many comparisons between the wording of the Irish Act and that legislation, and also provides pointers as to how the Irish Act might be interpreted having regard to the case law that has arisen from the Australian experience.

Unfortunately, despite the acknowledged desirability in all eight states/territories that the interpretation of legislation should be consistent, it appears that: 'more divergence than convergence is arising as time progresses'.[12] That diversity is likely to be enhanced rather than diminished by the introduction of extensive amending legislation in New South Wales and Queensland since that comment was made.

References

1 [1999] BLR 93 at page 97; (1999) 64 ConLR 1, [1999] 3 EGLR 7, [1999] CILL 1470, [1999] 37 EG 173, (1999) Times, 11 March, [1999] All ER (D) 143

2 Hansard HL volume 57

3 [2003] BLR 487; [2002] EWHC 2923 (TCC), 94 ConLR 81, [2003] All ER (D) 322 (Jan)

4 [2010] 3 IR 95; [2010] IESC 18

5 Tómas Kennedy-Grant QC, paper delivered to the Adjudication Society's sixth annual conference, 15 November 2007

6 *Balfour Beatty Construction Limited v Lambeth London Borough Council* [2002] EWHC 597 (TCC) (paragraph 29); 84 ConLR 1

7 Teena Zhang, Why national legislation is required for the effective operation of the security of payment scheme, 2009, 25 BCL 376

8 'Towards harmonisation of construction industry payment legislation: A consideration of the success afforded by the East and West Coast models in Australia', by Coggins Fenwick Elliott and Bell, *Australian Journal of Construction Economics and Building*, 10(3), 2010

9 Final report of the independent enquiry into construction industry insolvency in New South Wales, November 2012

10 Final report of the review of the discussion paper – Payment dispute resolution in the Queensland building and construction industry, May 2013

11 For example, 'Towards harmonisation of construction industry payment legislation' (note 8 above), and the Report on security of payment and adjudication in the Australian construction industry by the Society of Construction Law (Australia), June 2014

12 Teena Zhang 'Why national legislation is required for the effective operation of the security of payment scheme', 2009, 25 BCL 376 at 381

2 The scope of the Act

Construction contracts

The provisions of the Irish Act relate to construction contracts. A *construction contract* is defined by section 1(1) as meaning:

> *[a]n agreement (whether or not in writing) between an executing party and another party, where the executing party is engaged for any one or more of the following activities:*
>
> (a) *carrying out construction operations by the executing party;*
> (b) *arranging for the carrying out of construction operations by one or more other persons, whether under subcontract to the executing party or otherwise;*
> (c) *providing the executing party's own labour, or the labour of others, for the carrying out of construction operations.*

This definition largely mirrors the definition of the same words in section 104(1) of the UK Act of 1996, as amended. The definition is very wide and has been interpreted by the English courts liberally. Akenhead J in *Parkwood Leisure Limited v Laing O'Rourke Wales & West Limited*[1] found that a collateral warranty made between the main contractor and the operator of a swimming pool and leisure facility came within the definition on the basis that the contract to which it was collateral was one for the carrying out of construction operations and notwithstanding that the collateral warranty was retrospective in effect.

It is important to note that the judgment stated that not all collateral warranties given in construction projects can automatically be regarded as construction contracts for the purposes of the UK Act. The wording and factual background of each warranty will need to be reviewed to determine whether it is a contract for the carrying out of construction operations. However, a major consideration will be whether the warrantor: 'is undertaking to the beneficiary of the warranty to carry out construction operations'. It was also stated that if a professional is simply warranting that a past state of affairs has reached a certain level, quality or standard, then this may be a pointer against the warranty being a construction contract for the purposes of the UK Act.

Although the definition in the Irish Act is similar to that of the UK Act of 1996 as amended, it is not identical. The corresponding provision in the UK legislation (section 104(1)(a)) defines the relevant agreement in wider terms. It provides that the term construction contract means an agreement with a person for, inter alia: '(a) the carrying out of construction operations'. The Irish definition adds the words: 'by the executing party' and also provides that the agreement is one: 'where the executing party is engaged for the carrying out of construction operations'. The collateral warranty in the *Parkwood* case included a standard term whereby the contractor warranted that: 'it has carried out and shall carry out and complete the Works in accordance with the Contract'. The court held that by virtue of this definition, albeit the works were not being carried out for the beneficiary, the collateral warranty fell within the broad definition of a construction contract. It remains to be seen whether an Irish court would make a similar finding, given that the Irish Act appears to require that the agreement is one between the executing party and the party for whom the work is being carried out.

There is no uniform, or even similar definition for construction contracts internationally. The NSW Act as amended contains a very simple definition at section 4(1): 'Construction contract means a contract or other arrangement under which one party undertakes to carry out construction work, or to supply related goods or services, for another party'. That of course begs the question as to what is the definition of construction work. As with the Irish Act, that is a more complex issue.

Construction operations

The Irish Act does not refer to construction work in this context but to construction operations. It states:

> '[c]onstruction operations' means, subject to subsections (3) and (4), any activity associated with construction, including operations of any one or more of the following descriptions:

(a) construction, alteration, repair, maintenance, extension, demolition or dismantling of buildings, or structures forming, or to form, part of the land (whether permanent or not);

(b) construction, alteration, repair, maintenance, extension, demolition or dismantling of works forming, or to form, part of the land, including (without prejudice to the foregoing) walls, roadworks, power-lines, telecommunications apparatus, aircraft runways, docks and harbours, railways, inland waterways, pipe-lines, reservoirs, water-mains, wells, sewers, industrial plant and installations for purposes of land drainage, coast protection or defence;

(c) installation in any building or structure of fittings forming part of the land, including (without prejudice to the foregoing) systems of heating, lighting, air-conditioning, thermal insulation, ventilation, power supply,

       drainage, sanitation, water supply or fire protection, or security or com-
       munications systems;

(d) *external or internal cleaning of buildings and structures, so far as car-*
    ried out in the course of their construction, alteration, repair, extension
    or restoration;

(e) *operations which form an integral part of, or are preparatory to,*
    or are for rendering complete, such operations as are previously
    described in this subsection, including site clearance, earth-moving,
    excavation, tunnelling and boring, laying of foundations, erection,
    maintenance or dismantling of scaffolding, site restoration, landscap-
    ing and the provision of roadways and other access works and traffic
    management;

(f) *painting or decorating the internal or external surfaces of any building*
    or structure;

(g) *making, installing or repairing sculptures, murals and other artistic*
    works that are attached to real property.

This wording is identical to the UK Act of 1996 as amended with the following exceptions:

1   The introductory words in the UK Act of 1996 are: 'In this Part construc-
    tion operations means, subject as follows, operations of any of the following
    descriptions . . . '. The Irish definition is considerably wider in that it covers:
    'any activity associated with the construction, including operations' of the
    stated description.

2   In subsection (b) of the definition the Irish Act substitutes: 'telecommunica-
    tions apparatus' for '(electronic communications apparatus)'.

3   Subsection (g) of the definition is not included in the UK Act of 1996 defini-
    tion. Arguably, that addition was unnecessary given the width of the intro-
    ductory statement and the content of subsection (a).

There is surprising uniformity across the globe as to the definition of construction work/construction operations. Subsection (g) of the Irish definition seems to be unique. Other jurisdictions seem to have adhered closely to the wording of the UK Act of 1996.

    Whereas one might expect that shop-fitting works would be regarded as construction operations, it was held in *Gibson Lea v Makro*[2] that such works were not to be so considered where the units in question were not fixtures and were therefore not the construction of buildings or structures forming, or to form, part of the land.

    In *Potton Developments Limited v Thompson and Another*[3] it was held that self-contained bedroom 'pods' were not part of the land. The case concerned, inter alia, the distinction between fixtures and chattels and traced the origin of the modern law in the area to a statement by Mr Justice Blackburn in *Holland v Hodgson.*[4] In that case it was stated:

Perhaps the true rule is, that articles not otherwise attached to the land than by their own weight are not to be considered as part of the land, unless the circumstances are such as to shew that they were intended to be part of the land, the onus of shewing that they were so intended lying on those who assert that they have ceased to be chattels, and that, on the contrary, an article which is affixed to the land even slightly is to be considered as part of the land, unless the circumstances are such as to shew that it was intended all along to continue a chattel, the onus lying on those who contend that it is a chattel.

The court held that the units had not attached to the realty and were chattels and had always remained chattels. This was notwithstanding that the pods were attached by 30mm nails and ties to an exterior small brick wall. Judge Anthony Thompson QC (sitting as a Judge of the Chancery Division) noted:

The important and governing factors to my way of thinking are these. The units are clearly designed for both delivery and removal entire; what is more, they can be removed and indeed, as I saw on the video in the case of Forte Travel Lodge, can be removed without the units in any way being destroyed. The exterior embellishments may be taken off, but the units themselves are not in any way destroyed by their removal.

In *Palmers Limited v ABB Power Construction*[5] it was held that 'fabricating plant offsite' was 'to form' part of the land. In *Savoye & Another v Spicers Limited*[6] Akenhead J held that a contract to install industrial equipment incorporating a conveyer belt into a factory was a construction contract within the meaning of the Act. The equipment was firmly attached to the floor of the premises and for that reason would be regarded as a fixture rather than a fitting. The court pointed out, however, that this distinction of itself would not be determinative. The court also considered other factors such as whether the incorporation of the equipment would enhance the value of the premises and the degree of permanence of the structure.

In *Staveley Industries Plc v Odebrecht Oil & Gas Services Limited*[7] the court held that work involving the design and installation of electrical and telecommunications equipment for an oil and gas rig was not within the ambit of the UK Act of 1996, because the work to be carried out did not form part of the land. Interestingly, from an Irish perspective, the court came to the conclusion that there was no intention to include offshore installations within the UK Act of 1996, because the particular subsection of the UK Act of 1996 was derived from section 562(2) of the Income and Corporation Taxes Act 1988, which included a provision for offshore installations and this was deliberately omitted from the UK Act of 1996. This raises the question as to whether an Irish court, in interpreting Irish legislation, would be entitled or inclined to have regard to the fact that a provision included in the Irish legislation is identical to a subsection included in the earlier UK legislation and then have regard to the derivation of that legislation. It is highly unlikely that the Irish legislators were aware of this derivation and therefore formed a positive intention to exclude offshore installations.

Commencement

Section 12(2) states:

> *This Act applies in relation to construction contracts entered into after such day as the Minister may by order appoint.*

The Minister has by order appointed the 25th July 2016 as the date for the commencement of the Act. The Act therefore applies to all construction contracts entered into after that date. It can be difficult to ascertain exactly when a contract is entered into. In the *Atlas Ceiling*[8] case the court was asked to enforce the decision of an adjudicator by way of summary judgment. However, as there were disputes as to the date on which the relevant contract was entered into, the judge directed a hearing of the issue as to the jurisdiction of the adjudicator. Judge Thornton noted that the relevant part of the UK Act of 1996 applied only to construction contracts which were entered into after its commencement (1 May 1998). The parties had actually signed a document purporting to be a contract on 3 April 1998, but having reviewed all the facts of the case, the court came to the conclusion that a document executed on 12 April 1999 represented the parties' actual contract. This finding threw up an added complication that by reason of the document being executed on that date, which was after the commencement of the Act, it was retrospective in its effect to a period prior to the commencement of the legislation. The court found that for the purpose of the Act the date of execution of the contract, notwithstanding its retrospective effect, was the relevant date; therefore the adjudicator had jurisdiction.

Executing party

The Irish Act defines *executing party* as meaning:

(a) *where the parties to the construction contract are a contractor and the person for whom the contractor is doing work under the contract, the contractor, or*

(b) *where the parties to the construction contract are a contractor and a subcontractor or are 2 subcontractors, the subcontractor or whichever of the subcontractors agrees to execute work under the contract.*

None of the legislation operating elsewhere uses this expression and accordingly there is no case law referable to it.

Related services

Subsection 1(2) extends the definition of a construction contract to include related activities as follows:

(2) *In this Act references to a construction contract include an agreement, in relation to construction operations, to do work or provide services ancillary to the construction contract such as—*

    (a) *architectural, design, archaeological or surveying work,*
    (b) *engineering or project management services, or*
    (c) *advice on building, engineering, interior or exterior decoration or on the laying-out of landscape.*

This is the same as subsection 104(2) of the UK Act of 1996, except that the word 'archaeological' is omitted from subsection (a) of that provision; subsection (b) above is omitted in its entirety from the UK Act of 1996; and the words 'to provide' appear before the word 'advice' in subsection (c) above.

In *Fence Gate Limited v James R. Knowles Limited*[9] the TCC found that services provided by Knowles, the claims consultants, in relation to a construction arbitration were not to be regarded as arising from an agreement to provide services ancillary to the construction contract and did not fall within subsection 104(2).

There may be unintended gaps in the services listed in subsection 1(2) of the Irish Act. For instance, it is not clear whether work carried out by a quantity surveyor in preparing a bill of quantities would be included; it may or may not fall within 'surveying work', but in so far as it is undertaken before the project commences, it would hardly be regarded as a project management service. An example of a more holistic definition is afforded by the Northern Territory of Australia Act, which includes professional services and defines these as:

[s]ervices that are provided by a profession and that relate directly to construction work or to assessing its feasibility (whether or not it proceeds), including surveying, planning, costing, testing, architectural, design, plan drafting, engineering, quantity surveying and project management services, but not including accounting, financial or legal services.

(Section 7(2)(a))

Supply of goods

Subsections 1(3) and 1(4) provide that the Irish Act does not apply to supply or manufacture contracts unless the contract is also one for installation. The exclusion of supply contracts is not universal (international legislation more often than not includes supply contracts), and the logic for their exclusion is not apparent. The Australian Capital Territory Act, for example, applies to construction contracts and contracts for related goods and services. This is defined by section 8(1) as including contracts for: 'materials and components to form part of any building structure or work arising from construction work'.

Subsection 2(5): Legislation binding irrespective of the agreement between the parties

Subsection 2(5) is important. It states:

(5) *This Act applies to a construction contract whether or not—*

    (a) *the law of the State is otherwise the applicable law in relation to the construction contract, or*

    (b) *the parties to the construction contract purport to limit or exclude its application.*

This provision is obviously essential to the effectiveness of the Irish Act. It would have no teeth but for this provision. Curiously, earlier versions of the proposed legislation excluded this provision.

All of the relevant legislation of the countries that have adopted adjudication as a means of dispute resolution in the construction industry expressly prohibit contracting out of the legislation. A distinction, however, has to be drawn between the provisions which prevent contracting out of the legislation and provisions in such legislation that give the parties the choice of regulating matters through their contract or not. One of the main distinctions between the Australian East Coast and West Coast models was perceived to be the entitlement under the East Coast model to decide when and how progress payments were to be made through contract. The legislation did not seek to put any limit on the timeframe. The tendency now, however, is towards imposing the maximum time limit by statute, which will override the terms of contract, in all jurisdictions including the East Coast. For instance, the New South Wales Amendment Act of 2013 introduces an obligation to make a progress payment at the latest on the date occurring 15 business days after a payment claim is made.[10]

Subsection 2(5) may affect provisions in contracts seeking to apply foreign law. It would not be unusual, for instance, for a contract between a French main contractor and an Irish sub-contractor to provide, notwithstanding that the works are to be carried out in Ireland, that English law will apply. Such a provision is likely to be struck down in so far as it relates to disputes in respect of payment. This was the fate of such a clause relating to works to be carried out in New South Wales. Section 34 of the New South Wales 1999 Act prohibits parties from excluding the operation of that Act. In *Proactive Building Solutions v Mackenzie Kech*,[11] the Supreme Court of New South Wales struck down a clause as void in its entirety, because it provided that any disputes between the parties would be resolved in accordance with English law.

References

1 [2013] EWHC 2665 (TCC); 150 ConLR 93, [2013] 3 EGLR 6, [2013] BLR 589, [2013] All ER (D) 221 (Aug)

2 [2001] BLR 407; [2001] Lexis Citation 1688, [2001] All ER (D) 333 (Jul)

3 [1998] unreported; [1998] Lexis Citation 1947; [1998] NPC 49

4 [1872] L.R. 7 C.P. 328; 41 LJCP 146, 20 WR 990, [1861-73] All ER Rep 237, 26 LT 709

5 [1999] BLR 426; (1999) 68 ConLR 52, [1999] All ER (D) 1273

6 [2014] EWHC 4195 (TCC); HT-14-311, (Transcript)

7 [2001] 98(10) LSG 46; [2001] All ER (D) 359 (Feb)

8 *Atlas Ceiling & Partition Co Limited v Crowngate Estates (Cheltenham) Limited* [2000] TCC C.I.L.L. 1639 QBD (TCC); [2002] 18 Const. L.J. 49;

9 [2002] All ER (D) 37 (May); (2001) 84 ConLR 206, [2001] Lexis Citation 1933,

10 Section 3(1A) Amending Section 11 of the 1999 Act

11 [2013] NSW Supreme Court 1500

3   Exclusions from the scope of the Act

Subsections 2(1) – 2(4): Construction contracts not covered by the Irish Act

These are as follows:

- a contract to the value of less than €10,000;
- a contract in relation to a private dwelling only between a contractor and the person who occupies, or intends to occupy, the dwelling where the floor area is not greater than 200 square metres. A contract between a main contractor and sub-contractor in relation to such dwellings would be covered by the legislation. Section 106 of the UK Act of 1996 excludes the application of that Act to: 'A construction contract with a residential occupier';
- a contract of employment. This is a universal exclusion;
- a public private partnership contracts.

The value exclusion

The exclusion of contracts below a minimum value is unique to Ireland. Why or how this exclusion came about is not clear. The Regulatory Impact Analysis of the Construction Contract Bill published in September 2011 does offer some insight. At the time of publication, the Bill was proposing that the legislation would not apply to contracts involving a state entity where the value of the contract was less than €50,000, or, in relation to the private sector, where the value of the contract was less than €200,000. According to the analysis: 'it was considered that applying the legislation to contracts below these levels would place a disproportionate regulatory burden on the parties to the contracts'.[1] However, the analysis goes on to point out that, according to the available statistics of the UK experience, the majority of disputes availing of adjudication were in relation to contracts valued between £10,000 and £50,000.[2] One might infer, therefore, that the decision to exclude contracts with a value of less than €10,000 was made to avoid a disproportionate regulatory burden on the parties to the contracts. It is to be noted that no other country had considered it necessary to introduce such a safeguard, notwithstanding that amending legislation had been introduced in some. It is perhaps surprising, therefore, that the Irish legislature identified a regulatory burden that no other country with actual experience of adjudication had noted.

It has to be borne in mind that statutory adjudication is only an option. If parties do not wish to avail of it, they are not obliged to do so.[3] A party is unlikely to embark upon adjudication if the expense of the adjudication is disproportionate to the amount of the claim. However, adjudication may be an appropriate option where a debt is clearly due and the debtor is simply seeking to avoid payment. The fact that the debtor will almost inevitably in such circumstances have to pay the adjudicator's fees should discourage debtors from avoiding payment in such circumstances.

There are some quite interesting statistics available from the New South Wales Adjudication Research and Reporting Unit. The Annual Report for 2012/2013 shows that there were 672 adjudication applications in respect of which determinations were released. Of these, 225 related to claims of less than $10,000 (circa €7,000). The adjudicator's fees averaged about 18 per cent of the amount claimed, but the respondent on average was obliged to pay 97 per cent of the fees and the claimant only 3 per cent. This would suggest that claimants only pursued claims in this range where they had a clear entitlement.

Whereas the Irish Act excludes contracts below €10,000 in value, where standard contracts are used, it is likely that these will provide for adjudication, and that in those circumstances adjudication will be available even if the value is below €10,000.

In some jurisdictions (New Zealand for one) it is possible to register an adjudicator's decision as a judgment without having to make any formal application to the court. As the primary legislation does not provide for this, in Ireland an application will have to be made to the court before judgment can be entered. It might have been appropriate to have provided, where the value of an adjudicator's decision is relatively small, that it could be registered as a judgment without formal application being made to court.

The residential threshold

The exclusion of contracts between builders and house owners/occupiers is not unusual, but neither is it universal. Similar exclusions are included in legislation relating to the Australian Capital Territory, New South Wales, Queensland, Singapore, South Australia and Victoria. No such exclusion, however, is made in the legislation of the Northern Territories, Western Australia, Tasmania or New Zealand. Such contracts are excluded in Malaysia where the building concerned is less than four storeys high.

The exclusion, where it applies, reflects an intention that adjudication would apply primarily between businesses and not as between a business and a consumer. It is difficult to understand the logic for this. A contractor who is denied payment under a residential contract is as likely to be in need of cash flow as any other building contractor, and the availability of adjudication to such a contractor may be no less important than to any other contractor. If the legislation were particularly complex (as it is in some jurisdictions such as Victoria), there may be good reason not to subject consumers to it. However, the Irish Act is not complex and does not contain hidden traps whereby consumers may lose their entitlements unknowingly. If it were felt that consumers might be caught off guard by being subjected to adjudication, then rather than exclude residential contracts, it might have been more appropriate to include them, but to require that notices served under the Irish Act in relation to such contracts be more explicit, such as reciting the section of the Irish Act under which the notice is being served. The New Zealand Act requires the notice of adjudication, where the recipient is a residential occupier under a residential construction contract, to set out prominently a statement of the residential occupier's rights and obligations in the adjudication and a brief explanation of the adjudication process. Subject to that, residential contracts are not excluded.

While the Act does not apply in the circumstances stipulated to a contract between the residential owner and the main contractor, the Act does apply to any sub-contract made under the main contract. Therefore payments may become due and owing to sub-contractors in advance of payments becoming due and owing under the main contract. Under the Irish legislation there will be no opportunity for the main contractor to contract out of this obligation to its sub-contractors. Amending legislation introduced in New South Wales in 2013 expressly provides for the main contractor opting out of the otherwise obligatory time limits for payments to sub-contractors where the sub-contract is connected with an exempt residential contract.[4]

It is to be noted that the exemption applies where the contract relates "*only*" to a dwelling. If the contract therefore included a dwelling and an office or business facility the exclusion would not apply. Presumably the exclusion would apply, if the floor area is less than 200 square metres, to a holiday home given that the Act does not require that the residence be the sole or principal residence of the occupier.

Public private partnership arrangements

The Irish Act provides at section 2(3) that a contract between a state authority and its partner in a public private partnership arrangement (as those terms are defined in the State Authorities (Public Private Partnership Arrangements) Act 2002) is not to be regarded as a construction contract and that the Act will not therefore apply. This again is a provision which is unique to Ireland. Presumably, the rationale behind this is that entities that enter into such arrangements with state authorities tend to involve funding organisations and construction companies with huge resources, so that they have equal bargaining power to that of the contracting authority. This underlines the fact that this legislation has been introduced to avoid smaller sub-contractors being oppressed. The primary purpose is not to assist those who are perfectly capable of standing up for their own rights. The fact that large contractors, who may have resources of far greater value than those of their clients, have the benefit of the Act in all countries that have introduced such legislation, probably arises from the fact that it would be virtually impossible to legislate only for main contractors of a particular size. However, there is probably also a recognition involved in this that cash flow can be as much the life blood of the very large contractor as it is of the small contractor. The only difference very often is one of scale.

Other exclusions

As with legislation everywhere, the Irish Act provides that a contract of employment is not a construction contract. Otherwise the Irish Act does not have additional exclusions. The UK Act of 1996 provides for quite a number of additional exclusions. It also excludes contracts relating to the extraction of oil or natural gas, the extraction of minerals and works relating to power plants, water treatment plants and similar (section 105(2)(a) – (c)). Others have followed the UK precedent, but usually to a lesser degree. For instance, the Queensland Act excludes the drilling for, or extraction of, oil or natural gas and the extraction of minerals, but no more than that. These exclusions give rise to difficulty, because a contract may in part relate to construction operations covered by the legislation and in part to operations which are not. Furthermore, while the work undertaken by the main contractor may be excluded from the application of the legislation, the tasks undertaken by individual sub-contractors may not.

The *Palmers*[5] case involved an analysis of the meaning of 'construction operation' and 'construction contract'. *ABB* was a sub-sub-contractor for the assembly and installation of mechanical engineering work associated with a heat recovery steam generator boiler and *Palmers* were scaffolding sub-sub-sub-contractors. It was held that *Palmers'* scaffolding work was not taken outside the scope of construction work by section 105(2), although *ABB*'s sub-sub-contract work was. In noting the reasons for the exclusions under the UK Act of 1996, Judge Anthony Thornton QC stated:

> However, it is generally known that a limited number of contracting organisations representing specific sections of the construction and engineering industry persuaded the government to exclude the contracts of their members from the ambit of the Act. This was because these sections of the construction and engineering industry were already operating satisfactory contractual arrangements concerned with payment. This is the explanation for s 105(2) of the Act.

In considering the background to the UK Act of 1996, the court noted that the language of the subsections meant that a contractor may not necessarily be treated in the same way vis-à-vis their employer as a sub-contractor is to be treated by that contractor.

> Thus, it is perfectly possible, and within the statutory scheme, for a contractor's operations to fall outside the definition of a construction operation yet for a sub-contractor providing building, foundation or painting services for that contractor's work to come within the definition. This means that some sub-contractors are able to seek an adjudication and rely on the Act's

statutory restrictions on a contractor's powers of set-off whereas the same contractor is not able to seek adjudication under the relevant main contract next up the contractual chain nor to require the employer to use the Act's set-off procedures. This consideration counters ABB's argument that it would be unfortunate and contrary to the statutory scheme if ABB could be the subject of a set-off imposed by Stork and could not require an adjudication of any dispute with Stork as to the validity of that set-off whereas ABB might be unable to levy the same set-off against Palmers, because of the operation of Pt II of the Act, and could be the subject of an adjudication about that set-off at the behest of Palmers, despite Palmers being below ABB in the contractual chain.

In holding that all the activities that were being performed by *ABB* fall within the description of a construction operation the court noted:

> The ultimate conclusion as to whether the assembly process involved is to be considered as 'construction' is a question of law. Although, in popular speech, the word 'construction' is usually used in connection with building operations as opposed to engineering operations, the word clearly has a wider connotation when used in connection with many of the operations described in s 105(1). Thus, structures and works forming part of the land (which are not confined to buildings but are clearly intended to refer to all structures and works of whatever type) are linked with this word. Moreover, power-lines, telecommunications apparatus and industrial plant are expressly included within the definition of 'works' forming part of the land. Thus, it is clearly envisaged that the assembly and fixing to the land of industrial plant and similar features are included within the definition of construction operations and are also included in the definition of 'construction'.

The wording of section 2(1) of the Irish Act is such that in relation to residential developments excluded by the Irish Act, the exclusion will not apply to sub-contracts of any tier. Therefore, while the main contractor will not have any entitlements under the Act, it will still have to meet its obligations under the Act to its sub-contractors.

It is important to note that the standard contracts in use in the UK incorporate clauses providing for adjudication. The result is that contracts that are not covered by the legislation are subject to adjudication by the agreement of the parties. In *Picardi v Cuniberti & Cuniberti*[6] the court found that a contractual provision providing for adjudication was enforceable notwithstanding that the contract was not covered by the legislation by reason of the residential exception. In doing so, the court rejected the argument that the clause was an unfair term within the meaning of consumer protective legislation. No doubt the standard contracts operable in Ireland will be amended in due course to incorporate adjudication as a means of contract dispute resolution. Such a clause would have to provide detailed rules as to how the adjudication is to operate, or, alternatively, incorporate a standard procedure. Standard procedures are published by a number of institutions in the UK, including the Institution of Civil Engineers.

In *Nottingham Community Housing Association Limited v Powerminister Limited*[7] it was argued by the Plaintiff that a maintenance contract was not a construction contract in so far as it related to a domestic heating system on the basis that the heating system did not form part of the land. A similar point could be made in relation to the Irish Act as the wording of the definition of construction operations at paragraph (a) is identical. Mr Justice Dyson dismissed the argument noting:

> There is no justification for distinguishing between the repair and maintenance of the roof or cladding panels of a building and the repair and maintenance of the various systems described in paragraph (c). They are all part of a building. It is not a misuse of language to say that the maintenance of a building includes the maintenance of the various items mentioned in paragraph (c). It is worth emphasising the scope of those items. They include not only heating and air-conditioning systems, but also lighting, power supply, drainage, sanitation and water supply. These are all vital parts of a building, whose proper functioning is required if a building is to be fit for habitation.

In the UK development agreements, contracts of insurance, bonds and certain financing contracts are specifically excluded by the Minister's order.[8]

No contract/quantum meruit cases

It sometimes arises in arbitration or other dispute resolution processes availed of in relation to construction disputes, that one or other party claims that the works were carried out without any contract existing between the parties. This typically occurs where a letter of intent authorises work to be carried out up to a specific monetary value. Sometimes all of the terms are not yet agreed between the parties, and the purpose of the letter of intent is to allow the works to progress pending a final agreement on terms. Alternatively, all the terms may already be agreed, but it is specifically stated in the letter of intent that no contract will come into existence until a formal contract has been signed by the parties. In either case, if the works proceed beyond the limit of the letter of intent, it will often be found that the works were carried out without there being any contract between the parties.[9] Would the Irish Act apply to such a situation? It could be argued that the fact that the works proceed with the consent of both parties is evidence of an agreement of some nature. The essence, however, of such cases is that the executing party has not in fact been engaged to carry out the works. Often, the executing party will be under the impression that it has been so engaged, but in reality it has no contractual entitlements and must claim by way of equitable relief. It is submitted that in such cases adjudicators would not have jurisdiction (because the work did not arise out of a construction contract) and, accordingly, any determination by them as to the rights and entitlements of the parties would be invalid.

Works partly included by the legislation and partly excluded

Section 2(4) states:

> *Where a contract contains provisions in relation to activities other than those referred to in the definition of a construction contract and section 1(2), it is a construction contract only in so far as it relates to those activities.*

The UK Act of 1996 contains a similar provision at section 104(5), which provides: 'Where an agreement relates to construction operations, and other matters, this Part applies to it only in so far as it relates to construction operations'.

In the case of *Cleveland Bridge (UK) Limited v Whessoe-Volker Stevin JV*[10] the court was concerned with the detail of the exclusions set out at section 105(2) of the UK Act of 1996 relating to a sub-contract for works at the liquefied natural gas terminal at Milford Haven. These very particular provisions have not been carried through to the Irish Act, and it is unlikely that the Irish Courts will be troubled with the very technical issues of interpretation that arose in that case. However, in so far as contracts for professional services of a particular nature are included within the ambit of the Irish Act, there may well be circumstances where a contract for services will be in part covered by the Act, because it relates to construction operations, and in part excluded, because it does not.

The principles applied by the court, therefore, in the *Cleveland Bridge* case may have applicability to the Irish legislation. The court set out the exercise it had to engage in as follows:

> Logically there are three steps in this case in determining whether the Adjudicator had jurisdiction to determine the claim referred to her. First it has to be determined whether under s 105(1) the relevant work carried out by Cleveland Bridge comes within the definition of construction operations. The second related question is whether any part of the work and, if so, what part comes within the provisions of s 105(2) which are not construction operations within the meaning of the Act. Thirdly, to the extent that there are construction operations and works which are not construction operations, the question arises as to the effect of this on the jurisdiction of the Adjudicator and on the enforceability of her Decision.

The court held that part of the decision of the adjudicator was within her jurisdiction and part was outside her jurisdiction. The court considered whether it could or should save the adjudicator's decision by identifying the sums allowed within jurisdiction, but held that it would not be appropriate to do so in circumstances where there was no agreement between the parties on the issue. The issue of whether or not a court will sever the defective element of an adjudicator's decision arises in a number of contexts. The issue here arises out of the adjudicator assuming jurisdiction on disputes excluded from the ambit of the legislation. The issue also arises in the context of a failure on the part of the adjudicator to adhere to the requirements of natural justice, and can also arise in respect of the parties failing to take the measures required by the legislation in respect of notices or otherwise. As will be seen below at pages 40–43, whether or not the adjudicator's determination is enforced through severance depends largely on the approach taken by the court as to whether the legislation should be interpreted purposefully, that is that it should not be impeded by technicalities, or restrictively on the basis that the legislation tends to provide for rough justice, and that tendency should not be encouraged by interpreting the provisions liberally.

References

1 Page 21
2 Adjudicating Reporting Centre – Report 10 June 2010
3 The Collins Report recommended that statutory adjudication would be compulsory in New South Wales (page 371)
4 Section 11(1C), of the Building & Construction Industry Security of Payment Act 1999 as amended Act 93 of 2013, schedule 1(31)
5 *Palmers Limited v ABB Power Construction Limited* [1999] BLR 426; (1999) 68 ConLR 52, [1999] All ER (D) 1273
6 [2003] BLR 487; [2002] EWHC 2923 (TCC), 94 ConLR 81, [2003] All ER (D) 322 (Jan)
7 [2000] 75 ConLR 65, [2000] BLR 759, 16 Const LJ 449, [2000] All ER (D) 1045
8 Construction Contracts (England and Wales) Exclusion Order 1998 (SI 1998/648) as amended by SI 2004/696.
9 *British Steel v Cleveland Bridge* [1984] 1 All ER 504; 24 BLR 94; [1982] Com LR 54;
10 [2010] EWHC 1076 (TCC); 130 ConLR 159, [2010] NLJR 768, [2010] BLR 415, [2010] All ER (D) 206 (May)

4   Payment entitlements

Subsections 3(1), (3) and (4): Entitlement to progress payments

(1) *A construction contract shall provide for—*

    (a) *the amount of each interim payment to be made under the construction contract, and*

    (b) *the amount of the final payment to be made under the construction contract, or for an adequate mechanism for determining those amounts.*

(2) *. . . . . .*

(3) *The Schedule shall apply to a main contract if and to the extent that it does not make provision for the matters specified in subsections (1) and (2).*

(4) *The Schedule shall apply to a subcontract except to the extent that it makes provision which is more favourable to the executing party than that which would otherwise be made by the Schedule.*

It is to be noted that the Irish Act very clearly indicates that it relates not only to progress payments but also to the final payment due under the contract. Many of the non-European Acts refer to progress payments only leading to some confusion as to whether that expression could be interpreted to include the final payment due under the contract. Because of a subtle difference in wording between the NSW Act and the Singapore Act, which was in fact modelled on the NSW Act, the term progress payments was found in Singapore to include[1] the final payment but in New South Wales was found not to do so.[2]

Subsections 3(1), (3) and (4) require that a construction contract makes provision for interim payments and a final payment, and provide through a schedule for the timing of such payments where they are not specified. Subsections (3) and (4) make an important distinction between main contracts and sub-contracts. The parties to a main contract are free to agree the timing for making interim applications and the periods for payment on foot of interim applications and final payment applications. The Schedule provides for the relevant periods in default. In relation to sub-contracts, however, the Schedule will apply unless the terms as set out in the sub-contract are more favourable to the sub-contractor executing the works.

This distinction between the terms applying to main contracts and sub-contracts is unique to Ireland. Furthermore, the decision not to have a mandatory time limit in respect of progress payments in the case of main contractors is against the trend of international legislation such as the New South Wales 2013 Amending Act. However, it does go further than the UK Act of 1996, which has not been amended in this regard. The UK legislation provides that the parties are free to agree the amounts of the payments and the intervals at which, or circumstances in which, they become due.[3] This provision applies equally to main contracts and sub-contracts in the UK.

In a recent case in New South Wales,[4] the adjudicator in a challenged decision relied on the statutory 'right to be paid' in deciding in favour of a contractor who had failed to comply with a notice requirement in respect of a variation. The adjudicator, in finding for the contractor, concluded that the contract, which effectively contained a condition precedent notice provision, offended against Section 3 of the NSW Act which, according to the adjudicator, provides that a person who undertakes work is entitled to be paid for that work, and that the terms of the contract cannot provide a bar to that entitlement.

The decision was challenged on the basis that in misconstruing the provisions of the legislation the adjudicator had committed a fatal jurisdictional error. The court, in upholding the challenge and finding that the decision was void for jurisdictional error, observed that a fundamental matter for consideration by an adjudicator is the contract itself. The court approved the following passage

> The Act does not create a right to remuneration for construction work – that right is created by the construction contract. What the Act does is to create and regulate a right to obtain a progress payment. It is inherent in the concept of a progress payment that it be a payment on account of the amount ultimately due.[5]

The point is that there is no express or implied right to be paid either under the legislation or otherwise. That entitlement is to be ascertained from the terms of the contract.

Subsection 3(1) stipulates that every construction contract shall provide for the amount of each payment (both interim and final), or an adequate mechanism for determining same. Under subsection 3(2) the contract must state the date upon which each payment claim shall be made, or an adequate mechanism for determining the date on which such a claim may be made, and specify the interval between the payment claim date and the date on which that payment is due. If it fails to do so, then the Schedule applies. The relevant provision in the Schedule, in so far as it relates to progress payments, provides at paragraph 4 that the amount of an interim payment is to be the difference between:

(a) *the aggregate of the gross value (determined in accordance with the construction contract) of the work done under the construction contract at the payment claim date concerned together with any additional amounts*

> *in the interim payment under the construction contract, less any deductions from payment provided for by the construction contract, and*
>
> (b) *the aggregate amount of interim payments that have already been made at that payment claim date.*

Presumably, when referring to any 'additional amounts in the interim payment' the legislation has in mind payment for materials on site, but not yet incorporated in the work (where the contract allows for such a payment), the release of retention and the like. The Schedule goes on to state in relation to such payments:

> (5) *The aggregate of payments made under a construction contract shall not exceed—*
>
> (a) *the amount provided for in the construction contract as originally concluded, and*
>
> (b) *amounts provided for by any amendments to that contract agreed between the parties.*

The expression 'any amendments' in this context is difficult to understand. Is the expression seeking to capture something over and above what would normally be termed as variations? If so, and the contract provides for variations, would such variations be covered by 5(a)? Arguably not, because there would be no 'amount provided for' variations in the original contract sum. Perhaps 5(b) is intended to cover variations to the contract works, which were not within the scope of the original contract and require therefore the agreement of the parties by way of amendment? The answers to these questions are by no means clear.

It is no secret that those responsible for drafting the legislation in Ireland primarily had regard to the legislation in the UK. To a large extent, the Irish Act followed that legislation almost verbatim. Presumably, the draftsman was seeking to capture some element in this provision which was perceived to be omitted from paragraph 2(2) of the UK Scheme. Unfortunately, it is not clear what element this was. Had the definition in the UK Scheme proven anyway problematic, presumably it would have been amended when the opportunity presented itself through the amendment to the UK Scheme effected by Statutory Instrument No 2333 of 2011. On the face of it, therefore, the equivalent provision in the UK Scheme was not problematic, whereas the wording availed of in the Schedule to the Irish Act undoubtedly is.

The equivalent provision in the UK Scheme[6] providing for stage or periodic payments provides that these are to be the aggregate of the following:

> (a) an amount equal to the value of any work performed in accordance with the relevant construction contract during the period from the commencement of the contract to the end of the relevant period (excluding any amount calculated in accordance with sub-paragraph (b)),
>
> (b) where the contract provides for payment for materials, an amount equal to the value of any materials manufactured on site or brought onto site

for the purposes of the works during the period from the commencement of the contract to the end of the relevant period, and

(c) any other amount or sum which the contract specifies shall be payable during or in respect of the period from the commencement of the contract to the end of the relevant period.

The UK Scheme at Part II, paragraph 3 states that where the parties fail to provide: 'an adequate mechanism for determining' the amount due, the provisions of the Scheme will apply. The adequacy of a contractual mechanism for determining the value of an interim payment was considered in the case of *Maxi Construction Management Limited v Morton Rolls Limited*.[7] In that case the court concluded that whilst the construction contract provided a mechanism for determining the value of each payment (being the reaching of an agreement with the employer's agent) the mechanism was not adequate. The UK Scheme at paragraph 12 required that a claim by the payee would specify the basis upon which the claim was calculated. The mechanism provided for in the contract did not require this detail. In the circumstances, the mechanism failed to meet the requirements of the UK Act of 1996, and the terms of the UK Scheme for Construction Contracts were therefore applied.

The Irish legislation contains a similar default position in so far as the Schedule is to apply if the contract does not provide: 'an adequate mechanism for determining' the amount due (Section 3(3)).

Subsections 3(2) – (4): Payment claims

(2) *A construction contract shall provide for—*

    (a) *the payment claim date, or an adequate mechanism for determining the payment claim date, for each amount due under the construction contract, and*

    (b) *the period between the payment claim date for each such amount and the date on which the amount is so due.*

(3) *The Schedule shall apply to a main contract if and to the extent that it does not make provision for the matters specified in subsections (1) and (2).*

(4) *The Schedule shall apply to a subcontract except to the extent that it makes provision which is more favourable to the executing party than that which would otherwise be made by the Schedule.*

The intended combined effect of subsections 3(1) and 3(2) is that every construction contract will, either expressly or by statutory intervention, contain provisions under which the parties to that contract will know or be able to understand the value of each interim payment, when a claim in respect of each payment will be made, and when the claimed payment will be made. The effect of subsections 3(3) and 3(4) set out above is that the Schedule will only apply in the case of main contracts if the main contract fails to provide for payment claim dates. In the case of sub-contracts, however, the Schedule will apply unless the contract contains terms more favourable to the sub-contractor.

The Schedule at paragraph 1 sets out the payment claim dates applicable where the Schedule applies. These are:

a   *30 days after the commencement date of the construction contract;*
b   *30 days after that date and every 30 days thereafter up to the date of substantial completion;*
c   *30 days after the date of final completion.*

Under subsection 109(2) of the UK Act of 1996, the legislation recognises that the parties are: 'free to agree the amounts of the payments and the intervals at which or circumstance in which they become due'. According to subsection 109(3), where the parties fail in this respect: 'the relevant provisions of the Scheme for Construction Contracts apply'.

The UK Scheme is a detailed document, in two parts – the first dealing with adjudication. The second, entitled 'Payment' sets out, in full, the payment provisions to be implied into a construction contract in circumstances where the parties to that contract have failed to reach agreement in respect of either or both of the following matters:

a   the amount of any instalment or periodic payment;
b   the intervals at or circumstances in which such payment falls due.

The UK legislation makes no distinction here between main contractors and sub-contractors.

Subsection 110(1)(b) of the UK Act of 1996 requires that the final date for payment of any sum which becomes due be defined with certainty in the construction contract. In *Alstom Signalling Ltd v Jarvis Facilities Ltd*,[8] the sub-contract provided that payment would be made by the main contractor within seven days of its receipt of a certificate under its main contract with the employer. The court considered that this mechanism adequately identified a final date upon which the payment would fall due and therefore satisfied subsection 110(1)(b). (This would be no longer permitted under the UK Act of 2009.)[9]

As set out above, the equivalent Irish provision also requires certainty with regard to payment dates. Under subsection 3(2)(b) the construction contract must provide for the interval between the payment claim date and the payment due date, in default of which, by virtue of subsections (3) and (4) of Section 3, the period set out in the Schedule (30 days) will apply.

Subsections (3) and (4) deal with the default provisions – comprised in the Schedule. As indicated, the Irish Act draws a distinction between main contracts and sub-contracts, and it is anticipated that this distinction will, at the very least, lead to a necessity for careful planning on the part of main contractors in dealing with cash flow.

Under subsection (3), in the event that the construction contract, being a main contract, does not contain specific provisions (or adequate mechanisms where appropriate) for determining the timing and amount of payments to be made, the provisions of the Schedule will apply.

All of the standard forms of main contract in common use in Ireland provide mechanisms for determining the timing of payments and their value, and all therefore would appear to meet the statutory requirements sets out in subsections 3(1) and (2). These standard form contracts were all, however, drafted long before this legislation was even contemplated, and it will be interesting to see whether, if the question is asked, the Irish Courts will consider that such standard form mechanisms are 'adequate' within the context of the Irish Act.

Under subsection (4) the Schedule will apply to all sub-contracts, whether or not the parties make specific provision, or include adequate mechanisms for determining the timing and value of payments, unless the terms set out in the sub-contract in question are more favourable than those set out in the Schedule. Having regard to the terms of the Schedule, it is unlikely that more favourable terms would ever be widely offered to sub-contractors, and, in the circumstances, it is relatively safe to assume that the terms of the Schedule will be incorporated into most, if not all, sub-contracts.

It is clear, therefore, that whilst the parties to a main contract are, for the purposes of this legislation, relatively free to agree payment terms and mechanisms between themselves, the situation as between main contractors and sub-contractors is considerably different. In this context it is opportune to record that whilst most of the traditional forms of main contract in common use in Ireland are the product of negotiations between stakeholders and representative organisations, the main contract forms used by government and state agency employers are not negotiated forms and, judged by any objective standard, are heavily biased in favour of the employer. These contracts require main contractors to take on many risks, which would, under standard forms of contract, negotiated by stakeholders in the industry with the government agencies, traditionally remain with the employer. The requirement imposed by this legislation, that main contractors assume payment obligations to sub-contractors, which may not be matched by the terms of the main contract, both in the public and in the private sector, imposes another layer of risk on main contractors, which, even if it is capable of being priced, will not, in fact, in a competitive market, be covered by the tender sum.

This aspect of the legislation is unique to the Irish Act. Other countries either permit the parties to regulate the intervals for payment through their contracts, or alternatively impose a regime, which applies equally to main contracts and sub-contracts. Recent reforming legislation in New South Wales and Queensland has opted for the latter, rather than the former. Furthermore, the Collins Report recognised the need for a buffer period, providing for a longer period for the main contractor to pay the sub-contractor than for the employer to pay the main contractor:

The object of this buffer is to give additional time to the head contractor who, by reason of its position standing in the middle of the contractual relationship, will then be able to benefit from additional time to pay its sub-contractors thus improving its own 'cash flow' position . . . [10]

This is not to seek to introduce the 'pay when paid' proposition by the back door. The justification for this recommendation is the lag created by the fact that the clock begins to run at the bottom of the tiered structure when the progress payment claim from the sub-contractor is delivered to the head contractor. This means that the head contractor will be required to pay the sub-contractor's progress payment claim within 28 days of receiving that payment claim yet the head contractor will not be in a position to pass on those claims up the chain to the owner until it has received them from the sub-contractor. [11]

The decision to distinguish between main contracts and sub-contracts in Ireland curtails the ability of parties to those contracts to effect commercial arrangements that suit the individual parties. In most cases, main contractors, aware of and alert to the cash flow demands of the projects with which they are involved, will seek to ensure that the valuation and timing mechanisms in main contracts will be sufficient to meet the ongoing costs of the project. They will need to assiduously administer claims in accordance with those procedures to ensure that it is the ultimate beneficiary of the project that funds it, rather than the main contractor. Even with such management, however, this will not always be possible.

In some circumstances, operation of the main contract payment terms, combined with compliance with statutory obligations to sub-contractors, may place main contractors in precarious, if not impossible, positions. The following is illustrative:

- A contractor enters into a contract with an employer for the construction of a facility, incorporating specialist equipment, to be supplied and installed by a specialist sub-contractor.
- The overall contract value, to include works and all specialist equipment, is €10 million, of which €8 million will be due to the specialist.
- The main contract provides an adequate mechanism for determining the value of each interim payment, and a mechanism for determining each interim payment claim date. It also provides that the interval between the payment claim date and the payment due date will be 90 days.
- The main contract therefore meets all the requirements of section 3(1) and (2), and section 3(3) has no application – the provisions of the Schedule will not apply to the main contract.
- The main contractor in turn enters into a specialist sub-contract. The value of the sub-contract is €8 million. Notwithstanding the fact that the main contractor and the specialist are amenable to payment dates which correspond with those of the main contract and are prepared to make appropriate arrangements to accommodate such dates, they are prohibited, by statute, from contracting

out of the provisions of section 3(4), which states that 'the Schedule shall apply to a sub-contract'.

- The Schedule states that the first payment claim date shall be 30 days after commencement, and every 30 days thereafter. Payment is due 30 days after each payment claim date, with the final payment due 30 days after 'final completion'.
- The main contractor would be required, under statute, to make a payment to its sub- contractor 60 days before it is entitled, under the terms of the main contract, to payment from the employer.

Whilst the foregoing is a somewhat extreme illustration of the operation resulting from the statutory distinction, it is not beyond possibility and, to a greater or lesser extent, represents a potential risk to be faced by most main contractors under this legislation.

In other jurisdictions the same situation can arise for main contractors where the legislation applies as between the main contractor and the sub-contractor, but not as between the main contractor and the employer, for example, in respect of house construction contracts for residents, where such contracts are excluded. The main contractor is at risk in such circumstances of finding itself financially stretched. In Ireland, however, this is likely to be a far more frequent occurrence.

Paragraph 1 of the Schedule specifies that payment claim dates shall arise 30 days after the 'commencement date' of the contract, and every 30 days thereafter up to the date of 'substantial completion', with the final payment claim date 30 days after the date of 'final completion'.

The Irish Act offers no definitions of the terms 'commencement date', 'substantial completion' or 'final completion', and it must be assumed that these dates are to be derived from the construction contracts themselves.. Some commentators have suggested that the parties themselves may provide definitions for these terms in their contracts. This may only be permissible to the extent that the definition is consistent with the Act. To define for instance "*final completion*" as the end of the defects liability period may not be consistent with the intention of the Act.

The UK Scheme draws a distinction between the amount and timing of interim or periodic payments – dealt with in paragraphs 1, 2, 3 and 4 of Part Two of the UK Scheme – and the amount and timing of final payments, which are dealt with in paragraphs 3, 5 and 8. Under the UK Scheme, any periodic or stage payment is due on the later of 7 days after the 'relevant period' (28 days if none specified), or the date on which a claim is made. The final date for payment of a sum that is due, is 17 days from the due date.

The Schedule makes no such distinction in respect of the timing of payments; payment claim dates – whether interim or final, arise at 30-day intervals, and payment due dates – whether interim or final, arise 30 days after each payment claim date.

Whilst this paragraph goes a long way towards protecting the sub-contractor from abusive behaviour on the part of the main contractor, the legislation could have gone further. For instance, section 13(7) of the NSW Act, as amended by the 2013 Act, requires that a payment claim made by a main contractor of the employer must be accompanied by a statement that includes a declaration that all sub-contractors have been paid all sums that have become due and payable in relation to the main contract works. It is a criminal offence to knowingly provide such a statement that is false or misleading. Upon conviction, a main contractor may face a penalty of 22,000 Australian Dollars and/or three months' imprisonment.

These provisions may provide a further unintended advantage to the sub-contractor and problem for the main contractor in respect of retention. Paragraph 4 of the Schedule allows the main contractor to make deductions in accordance with the construction contract. Deductions for retention therefore can be made in respect of interim payments. However under section 3(1) the sub-contract must provide a mechanism for determining the amount of the final payment to be made and the payment claim date in respect of the final payment under paragraph 1(c) of the Schedule is to be not more than thirty days after the date of final completion. The term *"final completion"* is not defined – it may mean substantial or practical completion or it may mean completion of outstanding works or snags after substantial completion. It is unlikely that it would be defined as meaning the end of the defects liability period as final completion of all the works may be actually completed well in advance of that date. If the payment is to be a final payment, presumably it must include all retention – otherwise it would be a penultimate payment. It could be therefore that main contractors are obliged to release sub-contractors retention well in advance of the end of the defects liability period.

Paragraph 2 of the Schedule: Short-term contracts

Paragraph 2 of the Schedule makes provision for short-term contracts – where the duration of the works is agreed at less than 45 'consecutive days'. In such circumstances, the Schedule makes no provision for interim or stage payments, and the terms implied into such a contract sets the 'payment claim date' at 14 days following completion of the work under the contract itself.

Short contracts (less than 45 days' duration) are dealt with under paragraph 6 of the UK Scheme, which similarly provides for payment of the entire contract sum at the end of the project.

What happens if the parties get it wrong – and the contract period extends beyond the 45 day estimate contemplated in paragraph 2? If the parties have genuinely agreed that the duration of the contract works will be less than 45 days, it would appear that paragraph 2 would nonetheless apply and that an adjudicator would not have jurisdiction. In the case of *Tim Butler Contractors Limited v Merewood Homes Limited*,[12] although the parties referenced a programme to complete the works within four weeks, this was not a contract document and the adjudicator decided, because the parties had not agreed to a period of less than 45 days, that a similar provision did not apply. The Court upheld the decision on the basis that it was within the adjudicator's jurisdiction to decide the terms of the contract.

The term at paragraph 2 '45 consecutive days' could give rise to some ambiguity. Does paragraph 2 only refer to contracts requiring the works to be carried out on each consecutive day, in other words without a break for weekends or holidays? Presumably not, but the drafting might have been clearer. It is to be noted that the UK Act (section 109(1)) refers to a period of 45 days rather than 45 consecutive days.

Paragraph 3 of the Schedule: When payment is due

Paragraph 3 of the Schedule is important. It specifies that when the Schedule applies, payment must be made no later than 30 days after the payment claim date.

It should be noted that the Irish Act (in subsection 10(3)) draws a distinction between relevant periods for the service of notices under the Irish Act (which periods appear to exclude Saturdays, Sundays and Public Holidays) and other time periods mentioned. It must be concluded, therefore, that the time periods set out in paragraphs 1, 2 and 3 of the Schedule are intended to refer to calendar days – inclusive of weekends and public holidays.

Subsection 3(5): Pay when paid

(5) *Except after the occurrence of the circumstances specified in subsection (6), a provision in a construction contract is ineffective to the extent that it provides that payment of an amount due under the construction contract, or the timing of such a payment, is conditional on the making of a payment by a person who is not a party to the construction contract.*

Subsection 3(5) outlaws pay when paid provisions. It states that a provision in a construction contract is ineffective to the extent that it provides that payment of an amount due under the construction contract, or the timing of such a payment, is conditional on the making of a payment by a person who is not a party to the construction contract.

The UK Act of 2009 has gone further. It provides that a payment, which is conditional upon a certificate being issued under another contract, is not permissible for the purpose of calculating the sum due. Therefore, not only are pay when paid provisions outlawed, so too are pay when certified.[13]

Subsection 3(6): Circumstances where pay when paid permitted

(6) *The circumstances referred to in subsection (5) are:*

(a) *where the other person is a company other than an unregistered company—*

(i) *the commencement of its winding up pursuant to section 251 of the Companies Act 1963 where no declaration of solvency has been made under section 256 of the Companies Act 1963,*

      (ii) *the presentation of a petition to wind it up pursuant to section 213 of the Companies Act 1963,*

     (iii) *the appointment of a receiver in respect of any of its property or assets, or*

     (iv) *the presentation of a petition for the appointment of an examiner under the Companies (Amendment) Act 1990 in relation to it;*

(b) *where the other person is an unregistered company, the commencement of its winding up pursuant to section 345 of the Companies Act 1963;*

(c) *where the other person is an individual or partnership, the making of an application for adjudication under the Bankruptcy Act 1988 in relation to it;*

(d) *the making of a winding up or similar order by a court in relation to the other person;*

(e) *the occurrence of any event corresponding to those specified in this subsection under the law of any state to which Council Regulation (EC) No. 1346/2000 of 29 May 2000 on insolvency proceedings applies.*

This provision has the effect of incorporating into Irish law, in respect of construction contracts, the law in the UK in this regard. It provides in effect that any provision in a contract, which makes payment subject to the payer receiving payment from a third party, is ineffective unless that third party is insolvent.

The regime in the UK, whereby a pay when paid (or pay if paid) provision was valid if the third party was insolvent, was somewhat of an anomaly peculiar to the law in the UK. If it is intended that the risk to a party up the chain being paid is to be the risk of that party and not a party down the chain, there is no logical reason why that risk should be altered by the fact that a third party up the chain is insolvent. That precedent has not been followed elsewhere by legislation incorporating adjudication. Most of that legislation includes clauses of this nature, but do not provide for the exception set out in subsection 3(6). For instance the Australian Capital Territory Act at section 14(1) states: 'A pay when paid provision of a construction contract has no effect in relation to any payment for . . . construction work carried out or undertaken to be carried out under the contract'. It then goes on to define a 'pay when paid provision' as one:

(a) that makes the liability of one party (the first party) to pay money owing to another party (the second party) contingent on payment to the first party by a further party (the third party) of the whole or part of that money; or

(b) that makes the due date for payment of money owing by the first party to the second party dependent on the date on which payment of the whole or any part of that money is made to the first party by the third party; or

(c) that otherwise makes the liability to pay money owing, or the due date for payment of money owing, contingent or dependant on the operation of another contract.

The relevant provision in the Northern Territory Act is very straight-forward, and provides for no exception:

> A provision in a construction contract has no effect if it purports to make the liability of a party (party A) to pay an amount under the contract to another party contingent (whether directly or indirectly) on party A being paid an amount by another person (whether or not a party).[14]

Unless there is a specific provision in the contract providing for pay when paid and/or providing for the exception permitted by sub-section 3(6) of the Act, the paying party, such as a main contractor, will lose the benefit of the insolvency exception. Main contractors, and sub-contractors employing sub-sub-contractors, wishing to have the benefit of this exception should include in their contracts provisions to the effect that the sub-contractor or sub-sub-contractor will not be entitled to payment in the event of the party next above it in the chain becoming insolvent and the contractor or sub-contractor, as the case may be, not being paid as a consequence. This, however, would be to take advantage of an anomaly in the law, which should not exist. The main contractor has the opportunity to carry out a financial investigation into the affairs of the employer prior to entering into the contract. The risk of the employer becoming insolvent should be that of the main contractor and the main contractor alone.

References

1   *Tiong Seng Contractors (PTE) Limited v Chuan Lim Construction Pte Limited* [2007] SGHC 142, [2007] 4 SLR 364
2   *Jemzone Pty Limited v Trytan Pty Limited* [2002] NSWSC 395
3   Section 109(2)
4   *Lahey Construction Pty Ltd v Trident Civil Contracting PTY Ltd* [2013] NSWSC 176
5   Beneton J: *Roseville Bridge Marina Pty Ltd v Bellingham Marine Australia Pty Ltd* [2009] NSWSC 320
6   Part II paragraph 2(2)
7   [2001] Scot (D) 12/8
8   [2004] EWHC 1232 (TCC); 95 ConLR 55, [2004] All ER (D) 02 (Jun)
9   Section 110(1A) (6)
10 Page 355
11 Page 366
12 [2002] 18 Const. LJ 74; 12 April 2000, unreported
13 Section 110(1A)(b) of the UK Act of 1996 as amended by the 2009 Act
14 Section 12

5 Payment claim notices

Subsections 4(1) and (2): Content of payment claim notices

(1) *This section applies where, not later than 5 days after the payment claim date, an executing party to a construction contract delivers a payment claim notice relating to a payment claim to the other party or another person specified under the construction contract.*

(2) *A payment claim notice is a notice specifying—*

    (a) *the amount claimed (even if the amount is zero),*

    (b) *the period, stage of work or activity to which the payment claim relates,*

    (c) *the subject matter of the payment claim, and*

    (d) *the basis of the calculation of the amount claimed.*

Section 4 only applies in the event of the payment application being made not later than five days after the payment claim date, in other words the date provided for in the contract or the date provided for in the Schedule to the Irish Act as the case may be. It would appear that a claim made after five days is invalid. The Irish Act does not set out the consequences. Presumably, the contractor or sub-contractor is entitled to whatever payment the contract provides for by way of interim payment or, if there is no contractual provision, it will simply have to await the next payment claim date to pursue its entitlements.

The Irish Act is also silent as to the consequences of the notice served under subsection 4(2) not being in compliance with that subsection. It can be anticipated that if the Irish Act is interpreted with the same purposeful intent as was applied by the courts in the UK, then strict compliance will not be required. However, that is a matter of degree and, if the notice fails to comply to any substantial extent with the subsection, presumably it will be deemed not to be a proper notice and therefore not requiring a response from the paying party. The approach of the courts across the adjudication world is similar in this respect: the notice need not be perfect. In *George Development Limited v Canam Construction Limited*[1] the New Zealand Court endorsed the view expressed by the author, Richard Smellie: 'Although the (section 20(2)) requirements are mandatory, technical quibbles that they have not been complied with will probably receive scant attention'.[2]

The court also approved a finding of Windeyer J of the New South Wales Supreme Court in the case of *Hawkins Construction v Mac's Industrial Pipework*:[3]

> The arguments were that they contained the incorrect contract number and abbreviated the name of the Act under which the claim was made . . . As to the first, while the contract number may have been wrong in some cases, the claims did identify the work done. The second argument was that because the payment claims abbreviated the name of the Act, they did not fulfil a statutory requirement to name the Act. This argument might have had some weight in 1800. In 2001 an argument based on the absence of the word 'and' and the letters 'USTRY' has no merit. It should not have been put.

The court in the *George* case went on to observe:

> We acknowledge that the approach of this appellant was not as pedantic as those confronting Windeyer J, but the general observation that technical quibbles should not be allowed to vitiate a payment claim that substantively complies with the requirements of the Act is critical and needs to be weighed alongside the 'technocratic' interpretation advanced by *George*.
>
> (Paragraph 43)

Payment claim notices are a regular feature in all such legislation, as are payment responses. Essentially, the regime being put in place by such legislation is one whereby a sum is claimed by the executing party, the responding party must either accept it or reject it, in whole or in part, promptly, and any difference can then be referred to adjudication. There is a fair degree of consistency on an international basis in the provisions setting out what must be included in a payment claim notice. It must set out the work to which it relates, and it must set out how the sum claimed is calculated. There is, however, some variation as to the circumstances in which such a notice may be served and this may give rise to difficulties.

The Irish Act requires that the payment claim notice be given not later than five days after the payment claim date. That term is defined in section 1 as follows: 'Payment claim date, in relation to a construction contract, means the date when a payment claim in relation to an amount due under the construction contract is required to be made'.

Does this mean that the notice has to be given after the payment claim date, i.e. in the window period of five days, or can it be given prior to the payment claim date? A similar issue arose in relation to the NSW Act, which provides at Section 13(1): 'A person . . . who is or who claims to be entitled to a progress payment (the Claimant) may serve a payment claim on the person who, under the construction contract concerned, is or may be liable to make the payment'.

The Supreme Court of New South Wales found that notice served the day before the reference date would be premature and would be invalid, but for a contractual arrangement made by the parties in the particular case.[4] This decision was reached on the basis of a strict interpretation of the NSW Act, which required the application to be made on or from the reference date.

The Court in Victoria took a different view of an almost identical section contained within the Victorian Act. The court in that jurisdiction laid considerable emphasis on the fact that the person entitled to serve notice was not just the person who is entitled to payment but also a person who: 'claims to be entitled to a progress payment'.

As indicated, there is a strong movement towards harmonisation of legislation and of legislative interpretation in Australia. This would not just involve extensive amendments to the current legislation, it would also require a more uniform approach by the courts. It is clear that some courts favour a strict approach to the interpretation of the legislation, whilst others (as in *Seabay*[5]) favour a more purposeful approach.

The Irish legislation has adopted similar wording to that of the UK legislation. The Scheme, as amended by the 2011 legislation, requires the payer: 'not later than five days after the payment due date' to give a payment notice to the payee. Although the wording of the legislation is similar in this respect, these words are used in entirely different contexts. In Ireland, it is the executing party, or payee, who is restricted; in the UK the restriction applies to the payer. If the payer (or its authorised agent) fails to give notice of the amount payable within five days, any previous application made by the payee for payment will regulate the sum payable and, if no such application was made by the payee, the payee may, after the expiry of that five-day period, serve notice of the sum payable.

As to the detail to be set out in the payment claim notice, the following paragraphs from the judgment of Finkelstein J in *Protectavale Pty Limited v K2K Pty Limited*[6] relating to section 14(1) of the Victorian Act are often quoted with approval by commentators in the Southern Hemisphere:

> It is necessary to decide whether the invoice meets the requirements of s 14. The test is an objective one; that is, it must be clear from the terms of the document that it contains the required information: Walter Construction Group Ltd v CPL (Surry Hills) Pty Ltd [2003] NSWSC 266 at [82]. But the terms must be read in context. Payment claims are usually given and received by parties experienced in the building industry who are familiar with the particular construction contract, the history of the project and any issues which may have arisen between them regarding payment. Those matters are part of the context: Multiplex Constructions [2003] NSWSC 1140 at [76].
>
> The manner in which compliance with s 14 is tested is not overly demanding: Leighton Contractors Pty Ltd v Campbelltown Catholic Club Ltd [2003] NSWSC 1103 at [54] citing Hawkins Construction (Aust) Pty Ltd v Mac's Industrial Pipework Pty Ltd [2002] NSWCA 136 at [20] ('The requirements for a payment claim should not be approached in an unduly technical manner . . . As the words are used in relation to events occurring in the construction industry, they should be applied in a commonsense practical manner');

Multiplex Constructions [2003] NSWSC 1140 at [76] ('A payment claim and a payment schedule must be produced quickly; much that is contained therein in an abbreviated form which would be meaningless to the uninformed reader will be understood readily by the parties themselves'); Minimax Fire Fighting Systems Pty Ltd v Bremore Engineering (WA Pty Ltd) [2007] QSC 333 at [20] ('The Act emphasises speed and informality. Accordingly one should not approach the question whether a document satisfies the description of a payment schedule (or payment claim for that matter) from an unduly critical viewpoint').

Nonetheless a payment claim must be sufficiently detailed to enable the principal to understand the basis of the claim. If a reasonable principal is unable to ascertain with sufficient certainty the work to which the claim relates, they will not be able to provide a meaningful payment schedule. That is to say, a payment claim must put the principal in a position where they are able to decide whether to accept or reject the claim and, if the principal opts for the latter, to respond appropriately in a payment schedule: Nepean Engineering Pty Ltd v Total Process Services Pty Ltd (in liq) (2005) 64 NSWLR 462, 477; John Holland Pty Ltd v Cardno MBK (NSW) Pty Ltd [2004] NSWSC 258 at [18]–[21]. That is not an unreasonable price to pay to obtain the benefits of the statute.

(Paragraphs 10–12)

It is likely that this practical approach would be endorsed by the Irish courts. The penultimate sentence in this quote is probably the most important. A payment claim notice must have sufficient detail to enable the other party to respond to it in accordance with its rights and obligations under the statute.

Whereas the Irish Act is very restrictive in terms of the time within which a payment notice can be served, it is extremely broad in terms of the manner and time within which any dispute arising therefrom (or otherwise) may be referred to adjudication: 'A party to a construction contract has the right to refer for adjudication in accordance with this section any dispute relating to payment under the construction contract'.[7]

The effect of this latter provision will be dealt with later in Chapter 6, but in so far as that subsection allows a party to refer any dispute relating to payment to adjudication, and the following subsection 6(2) entitles it to do so: 'at any time', the five-day restriction imposed by subsection 4(1) may be of little consequence.

There is, in fact, a disconnect in the Irish legislation between the payment claim notice provisions and the entitlement to adjudication. There is no stated consequence for the failure to pay the amount claimed, even if it is not disputed, and adjudicators are not, in the authors' opinion, obliged, as they are in many other jurisdictions, to find that the undisputed amount claimed is due for payment under the contract (see page 46 below).

The words "*(even if the amount is zero)*" in sub-section 4(2)(a) would seem to be superfluous and without purpose. It is thought that they were included because they are included in the UK legislation, where they do have a purpose.

Severance generally and in relation to payment claim notices

What is the position if the payment claim notice is valid in part, but is invalid in other parts in that it fails to set out the requirements prescribed by subsection 4(3)? This issue arose in the Victorian case of *Gantley Pty Limited v Phoenix International Group Pty Limited.*[8] Applying first principles of common law as to severance, the court found:

> Severance in this case would operate to achieve the purpose and objects of the Act and would not operate to diminish the attainment of these goals. A respondent to a payment claim and an adjudicator, if appointed, should be able to assess the valid part of this progress claim which sufficiently describes the work for which payment is claimed, and provide a rational response or adjudication determination in respect of that part of the claim, and exclude from consideration that part of the claim which does not comply.

The issue of severance is more likely to arise under the Victorian Act than under any of the other statutes. This is because the Victorian Act uniquely expressly excludes from the process of adjudication specific claims, including those relating to variations of the construction contract, which are not claimable variations under the contract, latent conditions and time-related costs.

The courts in New South Wales[9] and in Queensland[10] took a different approach to that of the Victorian Courts on the basis, in both cases, that the relevant legislation required the adjudicator to make a finding of a single amount in respect of the progress payment. Once the adjudicator had made his decision, the court could not vary that decision by reducing the amount. In the *James Trowse* case the court observed: 'Save a slip rule, there is no mechanism available to sever any unlawful finding from an adjudicated amount, in particular a part of the adjudicated amount that is infected by jurisdictional error as found in this case'.[11]

Although the findings in these two cases (*Multiplex* and *James Trowse*) can be distinguished from the Victorian precedent on the basis of the wording of the legislation, the issue may equally be one of judicial approach. It may be that the Victorian Supreme Court took a more purposeful approach to the legislation, whilst the courts in New South Wales and Queensland found that severance could not be applied on the strict wording of the legislation.

The philosophy that underlies the strict approach to interpretation of the legislation was summarised by McDougal J as follows in relation to the New South Wales legislation:

> The Security of Payment Act gives very valuable, and commercially important, advantages to builders and sub-contractors. At each stage of the regime for enforcement of the statutory right to progress payments, the Security of Payment Act lays down clear specifications of time and other requirements to be observed. It is not difficult to understand that the availability of those rights should depend on strict observance of the statutory requirements that are involved in their creation.[12]

The issue of severance in the context of payment claim notices only arises if the requirements of the legislation in respect of such notices are to be regarded as conditions precedent. Whether they are or not may itself depend upon whether the court takes a liberal or restrictive approach to interpretation. If the particular provision is found to be a condition precedent, the question of whether or not the court should enforce an unaffected element of the adjudicator's decision may also depend on whether the court takes a purposeful or restrictive approach and, whatever approach is adopted, it is likely to be consistently applied irrespective of whether the adjudicator's decision is defective, because of lack of jurisdiction, failure to apply natural justice or otherwise.

The Technology and Construction Court initially arrived at a similar conclusion regarding severance in respect of an adjudicator's decision to that of the New South Wales and Queensland Courts. Akenhead J in *Cantillon Limited v Urvasco Limited*[13] stated obiter that where a single dispute is referred to adjudication and the adjudicator's decision is not enforceable because of a breach of jurisdiction or natural justice, the court cannot sever the element attributable to the breach and enforce judgment for the remainder. The court considered that it was not for it to decide how the adjudicator should have distributed the unaffected monies as between the parties. In this context, it is important to note that a single dispute (and the entitlement under Section 6 of the Irish Act is to refer a dispute rather than disputes to adjudication) will normally capture a large number of individual claim items. This fact does not constitute a multiple of disputes. The referral will only constitute two or more disputes if it, for instance, relates to different progress payments, and even that difficulty should be capable of being overcome through appropriate wording. Accordingly, nearly all referrals relate to a single dispute.

The legislation in the UK does not require the adjudicator to make a decision as to a specific amount, as is the case in New South Wales. The court therefore was not as restrained in terms of principle as was the Supreme Court in New South Wales. The logic of the position adopted by the TCC Court at that time was summarised in the decision of HHJ Seymour QC in *RSL (South West) Limited v Stansell Limited*[14] at paragraph 37 (and quoted in the *Cantillon* judgment at paragraph 61):

> The decision of an adjudicator is not like an award of an arbitrator or the judgment of a court and directly enforceable. It is enforceable at all simply because by their contract the parties have agreed to comply with it or to give effect to it. This is so whether the parties have expressly agreed in their contract to a procedure for adjudication, as in the present case, or whether the relevant provisions of The Scheme for Construction Contracts have been implied into their contract by virtue of the provisions of the Housing Grants, Construction and Regeneration Act 1996, section 114(4). What, on proper construction of clause 38A.7 of the sub-contract, the parties agreed to do in the present case, in my judgment, was to be bound and to comply with the decision of an adjudicator in relation to any dispute identified in a notice of adjudication which had been referred to him and which he had

determined . . . It is obviously conceptually possible for parties to an adjudication procedure to agree to be bound by, and to give effect to, not only the decision on the dispute referred but also any decision on a constituent element in the eventual overall total, or any process of reasoning adopted in the course of reaching a conclusion on the overall dispute. Whether that has happened in any particular case will depend upon the proper construction of the relevant contract, but it has not happened in the present case. *Thus once the decision as to the total amount to be paid has been successfully attacked, it cannot be said that any other amount has been determined by (the adjudicator) to be due in a way which is binding upon Stansell* [emphasis added].

Although the reasoning in *Cantillon v Urvasco* has been applied on a number of occasions, there is a sense of unease regarding the extent of its applicability. Judge Akenhead himself in *Pilon Limited v Breyer Group Plc*[15] suggested that the issue may have to be reviewed by the court. In a footnote to the ninth edition of Keating[16] the authors suggest that the court may be able to sever an adjudicator's decision in cases of breach of natural justice: 'but whether that is possible will depend, generally, on the nature of the breach of natural justice and whether there is more than one dispute'. The decision in the *Cantillon* case does seem harsh and, furthermore, not in the spirit of the legislation. Where the adjudicator's error is associated with a particular sum or where a maximum figure attributable to the error can be ascertained, a purposeful interpretation of the legislation would suggest that the remainder of the decision should be enforced. In *Quartzelec Limited v Honeywell Control System Limited*[17] the court found that the adjudicator had failed to consider a defence worth £36,500 when making a decision in favour of the contractor in the sum of £135,000 in circumstances where he was bound to consider that defence. The court, following the reasoning in *Cantillon*, felt that it could not enforce the balance of the decision.

More recently the tendency of the Technology & Construction Court has been to permit severance where this can be done without having to make assumptions as to what was intended by the adjudicator. For example in *Beck Interiors v UK Flooring Contractors Limited*[18] Akenhead J enforced that part of the adjudicator's decision that related to a dispute which had crystallised but refused to enforce the other part of the decision. The adjudicator had directed the respondent to pay the adjudicator's fees and expenses but the Court refused to enforce any aspect of that decision on the basis that the Court could not presume how the adjudicator would have dealt with the issue of his fees and expenses had the adjudicator known that part of the dispute was outside of his jurisdiction. In a recent address to the Adjudication Society in Ireland Mr Justice Stewart-Edwards suggested that the current policy of the court is to permit severance, albeit cautiously, but went on to caution that the Court of Appeal might take a different approach[19].

It is submitted that the more liberal and purposeful approach to the issue of severance is to be preferred over the restrictive approach demonstrated above by reference to the earlier cases having regard to the overall purpose of the legislation. It is difficult for the courts to swallow the bitter pill that the legislation inevitably involves an element of rough justice, and there may be a tendency to rein in the possibility of rough justice by applying a restrictive approach. Given, however, that the possibility for rough justice, or indeed injustice leading to an undeserved payment occurring is not connected logically to the issue of severance, but arises by the nature of the legislative regime, there does not appear to be much purpose to applying the brakes in this manner. In fact, doing so is likely to cause injustice rather than prevent it, as is apparent from cases such as *Quartzelec* referred to above.

Subsection 4(3): Response to payment claim notices

4(3) *If the other party or specified person referred to in subsection (1) contests that the amount is due and payable, then the other party or specified person—*

  (a) *shall deliver a response to the payment claim notice to the executing party, not later than 21 days after the payment claim date, specifying—*

   (i) *the amount proposed to be paid,*

   (ii) *the reason or reasons for the difference between the amount in the payment claim notice and the amount referred to in subparagraph (i), and*

   (iii) *the basis on which the amount referred to in subparagraph (i) is calculated, and*

  (b) *if the matter has not been settled by the day on which the amount is due, shall pay the amount referred to in paragraph (a) to the executing party not later than on that day.*

Subsection 4(3) requires that the paying party, or if applicable, the third party charged with determining the amount due under the contract, to respond within 21 days after the payment claim date if they contest that the amount claimed is payable. If therefore the notice is received 5 days after the payment claim date, the paying party must respond within 16 days. Nonetheless, the time allowed under the Irish Act is quite generous. With the exception of the Singapore Act, all other jurisdictions require a quicker response.

In so far as this subsection sets out a requirement for the paying party to respond to the payment notice and to explain the reason why the full amount claimed is not to be paid, this repeats a requirement which is common to all such legislation. The Irish legislation requires the other party or specified person to specify in this response the reasons for the difference. The legislation elsewhere sometimes requires the paying party to 'indicate' rather than 'specify' the reasons for the difference. The word 'specify' may be taken to infer a more specific or exact response is required. In considering the word 'indicate' in the context of the New Zealand Act in the case of *Solidcrete Technology Limited v First Pacific Investments Limited*,[20] the court set out the Shorter Oxford Dictionary meaning of the word 'indicate' as being: 'to express briefly, lightly or without development; to give an indication'. The word 'specify' is defined in the Collins English online dictionary as meaning: '1. To refer to or state specifically; 2. To state as a condition; 3. to state or include in the specification of'.

The word used in the NSW Act at that time was also 'indicate'. In the *Solidcrete* case the court referred with approval to the judgment in the New South Wales case of *Multiplex Construction Pty Limited v Luikens*[21] where Palmer J stated:

> Section 14(3) of the Act, in requiring a respondent to 'indicate' its reasons for withholding payment, does not require the payment schedule to give full particulars of those reasons. The use of the word 'indicate' rather than 'state', 'specify' or 'set out', conveys an impression that some want of precision and particularity is permissible as long as the essence of the 'reason' for withholding the payment is made known sufficiently to enable the claimant to make a decision whether or not to pursue the claim and to understand the nature of the case it will have to meet in an adjudication.

What is clear, whether the requirement is to indicate the difference or specify the difference, is that general descriptions of a potential counterclaim will not suffice, as was clearly demonstrated by another New Zealand High Court case:[22]

> An assertion that remedial work is required at a cost which would exceed the payment claim could never constitute a valid reason either for the difference between the scheduled amount and the amount claimed or for withholding payment. General and unspecified allegations of defective workmanship are insufficient unless quantified within a reduction for the claimed cost of remedial work.
>
> (Harrison J)

Where the Irish Act deviates from others is in the following respects:

1   The Irish Act does not set out the consequences of a failure on the part of the paying party to deliver a response within the required time.
2   The Irish Act does not state that in the absence of a response the amount claimed in the payment claim notice is payable.

Whereas international practice does vary as to the extent to which a party who fails to serve a response document to the claim notice is entitled to dispute the amount of the claim in subsequent adjudication proceedings, the legislation elsewhere does at least state the consequences of not providing a notice.

Under the UK Act of 1996, whether or not a responding notice was served, what was to be paid to the claimant was the sum 'due under the contract' as opposed to the amount claimed in the payment notice. This afforded the paying party an opportunity to dispute in adjudication the amount claimed, on the basis that it was not properly calculated in accordance with the contract. The strong tendency in recent years has been to move away from that position and to legislate for payment of the amount claimed, irrespective of whether it is properly calculated under the contract, if the paying party fails to serve a responding/withholding notice. The UK has followed that tendency through the introduction of the UK Act of 2009. It provides that the claimed ('notified') amount is due for payment where the paying party fails to serve the required notice. Whether the Irish Legislature has opted to follow the regime that used to apply under the 1996 Act, or that now applicable under the 2009 Act is not clear. On the one hand, the Act does require a party wishing to contest the claim to serve a responding notice. On the other hand the Act does not say that the amount claimed becomes due if the paying party fails to respond.

New Zealand has operated a regime similar to that applied by the UK Act of 2009 since the introduction of its Construction Contracts Act in 2002. It provides for a payment claim followed by a payment schedule. If a payment schedule is not served, the claimed amount becomes due. Section 23(2)(a) is absolutely clear as to the consequences. It provides that in the event of a payment schedule not being served, the consequences are that the payee:

May recover from the payer, as a debt due to the payee, in any Court—

    (i)   the unpaid portion of the claimed amount; and
    (ii) the actual and reasonable costs of recovery awarded against the payer by that Court.

While most legislation provides for the amount set out in a payment claim becoming payable unless it is disputed in a response notice, the Malaysian Act provides at subsection 6(4) that if a party fails to respond to a payment claim, it is deemed to have disputed the entire payment claim. However, the Malaysian Act does not provide a regime for progress payments, and is not therefore concerned with progress payments as such, and this may account for that peculiarity.

The most recent legislation dealing with adjudication is the amending legislation of the Queensland Building and Construction Industry Payments Act 2004, effective from 15 December 2014. The question of whether or not a respondent in an adjudication should be entitled to rely on defences not included in its response to the claim notice, was carefully considered in the Wallace Report. As indicated above on page 6 the Wallace Report recommended, and the legislation introduces, a distinction between standard claims and composite claims, the latter involving claims in excess of $750,000, latent defects or time-related claims. The legislation ties the respondent to its response document in so far as a standard adjudication is concerned, but a respondent facing an adjudication under the composite scheme is entitled to raise reasons for refusing all or part of the claimed amount, which had not been raised in the response document (the 'payment schedule').

It should be noted that under section 21 of the Queensland Act, before embarking upon adjudication, a claimant must give a respondent, who has failed to serve a payment schedule in respect of the payment claim, a second opportunity to provide a payment schedule (response) so that in effect the paying party has a second opportunity to set out its defence to a payment claim. This requirement applied even before the 2004 Act was amended and is not therefore applicable only to complex claims under the composite scheme. A similar requirement arises under the NSW Act at section 17(2). However, if a respondent fails to deliver a payment schedule in response to the payment claim notice or in response to the second invitation under section 17(2), the respondent is virtually excluded from making submissions in the adjudication. This is because section 22(2) sets out the only matters adjudicators are to consider in arriving at their decision and, in so far as they are entitled to consider submissions made by the respondent, it can only be submissions that have been made by the respondent in support of its payment schedule.

In so far as the Irish Act does not state that in default the amount claimed is due and in so far as the purpose of the notice is to set out what is due under the contract, it is submitted that the rights and obligations of the parties under Irish law will be similar to the entitlements prevailing in the UK prior to the introduction of the UK Act of 2009. Section 4(3) states what the paying party must do if it contests that the amount set out in the payment claim notice is due and payable. It can be argued that this infers that the sum claimed is payable unless the paying party responds accordingly. In the absence of clear words indicating that this is the case, as is the case in other jurisdictions, it is submitted that this would not be the preferred interpretation. Under the standard forms of contract the amount due is calculated by reference to the value of the work and in default the Schedule to the Act contains a similar provision. It is submitted that the wording of Section 4(3) is not sufficiently robust to substitute the amount claimed for the amount due simply because the paying party does not initially contest the amount claimed. The default situation that has been brought about in the UK where the paying party fails to serve a payment notice or a pay less notice has caused considerable difficulty for the Courts in that jurisdiction[23]. Colloquially adjudications arising on foot of such defaults have become known as "*smash and grab*" adjudications[24]. If the Irish legislation was interpreted to give rise to similar default entitlements, this may give rise to even greater constitutional difficulties.

The position in the UK prior to the introduction of the 2009 Act

In the early days of adjudication in the UK, it was thought that if the paying party failed to provide a withholding notice within the time required, then the full amount claimed was due. This view appeared to be supported by the courts in *Whiteways Contractors (Sussex) Limited v Impresa Castelli Construction UK Limited*[25] and *Millers Specialist Joinery Company Limited v Nobles Construction Limited.*[26] However, later decisions such as *SL Timber Systems Limited v Carillion Construction Limited*[27] and *Watkin Jones & Son Limited v Lidl UK GmBH*[28] have established that this was not always the case. If the amount had been certified for payment either under a main contract or under a sub-contract, the service of the withholding notice was essential (*Clark Contracts v Burrell*).[29] Furthermore, a claim for set-off or counterclaim had to be the subject matter of a withholding notice. However, if the claim was simply by way of application or invoice, it would appear that the paying party was entitled to dispute it in the adjudication on grounds such as the claim was exaggerated, or was not based upon agreed rates, or was based upon claims to variations which were not in fact variations. The justification for this approach in the UK arose out of the wording of Section 111 of the UK Act of 1996 which provided at subsection (1): 'A party to a construction contract may not withhold payment after the final date for payment of a sum due under the contract unless he has given an effective notice of intention to withhold payment'. Lord MacFadyen made the point in *SL Timber Systems* as follows:

> It is no doubt right, as the adjudicator pointed out, that, if the section did require a notice of intention to withhold payment as the foundation for a dispute as to whether the sum claimed was due under the contract, it would be relatively straightforward for the party disputing the claim to give such a notice. But that consideration does not, in my view, justify ignoring the fact that the section is expressed as applying to the case where an attempt is made to withhold a sum due under the contract, and not as applying to an attempt to dispute that the sum claimed is due under the contract.[30]

In *Rupert Morgan Building Services (LLC) Limited v Jervis & Another*,[31] the claimant builders undertook building work for the defendant clients under a standard form of contract. The contract provided that payments were to be made to the builders only in accordance with the architect's certificate, and that the clients were to pay the amount certified within a specified period, subject to any deductions and set-off due under the contract. The clients disputed that part of the certified sum was due, but did not issue a withholding notice under the UK Act of 1996.

It was held by the Court of Appeal that where an architect had issued an interim certificate under that contract, the client was required to pay the certified sum unless a withholding notice was given.

In the *Melville Dundas*[32] case the employer was entitled, even where no withholding notice was given, to rely on a subsequent event, which entitled it to determine the contract and to rely on a 'no payment' provision.

In that case, a contractor (then in receivership), sued the employer under a construction contract for an interim payment, which had been applied for just before the appointment of the receiver. The employer terminated the contractor's employment under the contract by reason of the appointment of the receiver. The contract incorporated the conditions of the JCT Standard Form of Building Contract with Contractor's Design (1998 edition). In holding for the employer and allowing the appeal, Lord Hoffman in the House of Lords stated:

> A provision such as clause 27.6.5.1, which gives the employer a limited right to retain funds by way of security for his cross-claims, seems to me a reasonable compromise between discouraging employers from retaining interim payments against the possibility that a contractor who is performing the contract might become insolvent at some future date (which may well be self-fulfilling) and allowing the interim payment system to be used for a purpose for which it was never intended, namely to improve the position of an insolvent contractor's secured or unsecured creditors against the employer.

Conclusion

If the Irish courts follow the precedent set by the courts in the UK as to the amount payable in default of a paying party's response notice, it is likely that the paying party will be entitled to argue in adjudication that the amount set out in the claim notice is not payable, because it has not been properly calculated in accordance with the contract and is therefore not due under the contract. The paying party, however, will not be entitled to raise counterclaims or cross claims arising under the contract or under other contracts.

Subsection 4(4): Claims and cross claims by the respondent

Subsection 4(4) states:

> *Where a reason for the different amount in the response is attributable to a claim for loss or damage arising from an alleged breach of any contractual or other obligation of the executing party (under the construction contract or otherwise), or any other claim that the other person alleges against the executing party, the response shall also specify—*
>
> (a) *when the loss was incurred or the damage occurred, or how the other claim arose,*
> (b) *the particulars of the loss, damage or claim, and*
> (c) *the portion of the difference that is attributable to each such particular.*

This is a curious provision in that it provides for the paying party to make deductions for monies alleged to be due to it for reasons unconnected with the construction contract. The wording of some contracts would prohibit such deductions being made. Presumably such a contractual provision would apply notwithstanding this subsection and subsection 4(5) which states:

> *The rights and obligations conferred or imposed by this section are additional to any conferred or imposed by the terms of the construction contract.*

It is thought that subsection 4(4) does not confer a right to make such deductions if the contract itself does not allow such deductions to be made, but merely provides what the paying party must do to enforce such a right, if it exists.

All relevant legislation internationally provides for the recipient of the payment claim to respond within a specified time and, if the claim is not accepted, set out the reasons why. The period of 21 days (from the payment claim date) allowed in the Irish Act for a response is unusually long, but the details required are more exacting than most. For example the NSW Act requires a response within ten business days and, where the amount as set out in the response as being payable is less than the amount of the claim, the respondent: 'must indicate why the scheduled amount is less and (if it is less because the respondent is withholding payment for any reason) the respondent's reasons for withholding payment'.[33] This is considerably less onerous than the response required by the Irish Act.

References

1 [2006] 1 NZLR 177, [2005] 18 PRNZ 84, CA 244/04, 12/4/05; [2005] NZCA 84
2 Progress Payments and Adjudication (2003), page 31
3 [2001] NSWSC 815
4 *Walter Construction Group Pty Limited v CPL (Surrey Hills) Pty Limited* [2003] NSWSC 266
5 *Seabay Properties Pty Ltd v Galvin Construction Pty Ltd & Anor* [2011] VSC 183
6 [2008] FCA 1248
7 Section 6(1)
8 [2010] VSC 106
9 *Multiplex Construction Pty Limited v Luikens & Another* [2003] NSWSC 1140
10 *James Trowse Construction Pty Limited v ASAP Plasterers Pty Limited and Others* [2011] QSC 145
11 Paragraph 59
12 *Chase Oyster Bar Pty Limited v Hamo Industries Pty Limited* [2010] NSWCA 190, paragraph 209
13 [2008] EWHC 282 (TCC); [2008] BLR 250; 117 ConLR 1, [2008] All ER (D) 406 (Feb)
14 [2003] EWHC 1390 (TCC)
15 [2010] EWHC 837 (TCC), 130 ConLR 90, [2010] BLR 452, [2011] Bus LR D42, [2010] All ER (D) 197 (Apr)
16 Page 722, Keating on Construction Contracts, ninth edition
17 [2008] EWHC 3315 (TCC); [2009] BLR 328
18 [2012] EWHC 1808 (TCC); [2012] BLR 417, [2012] All ER (D) 31 (Jul)
19 Adjudication Society Conference, Croke Park, 18th May 2016
20 [2005] DCR 769; CIV-2005-004-224
21 [2003] NSWSC 1140
22 *Metalcraft Industries Limited v Christie* Civ [2006] – 488–645
23 *Galliford Try Building Limited v Estura Limited* [2015] EWHC 412 (TCC) ; 159 ConLR 10, [2015] BLR 321, [2015] All ER (D) 01 (Mar)
24 *CG Group Limited v Breyer Group Plc* [2013] EWHC 2722 (TCC); 150 ConLR 1, [2013] BLR 575, [2013] All ER (D) 73 (Oct)
25 [2000] 16 Const LJ 453; (2000) 75 ConLR 92, [2000] All ER (D) 1171
26 [2001] TCC 64/00
27 (2001) 85 ConLR 79, [2000] BLR 516, 2001 SCLR 935, 2001 Scot (D) 45/6
28 [2002] EWHC 183 (TCC); 86 ConLR 155, [2002] All ER (D) 340 (Feb)
29 [2002] SLT 103; 2002 Scot (D) 26/4
30 Paragraph 22 of the judgment
31 [2004] 1 All ER 529; [2003] EWCA Civ 1563, [2004] 1 WLR 1867, 91 ConLR 81, [2003] NLJR 1761, [2004] BLR 18, (2003) Times, 26 November, [2003] All ER (D) 153 (Nov)
32 *Melville Dundas Limited (in Receivership) & Others v George Wimpey UK Limited & Others* [2007] BLR 257; [2007] UKHL 18, [2007] 3 All ER 889, [2007] 1 WLR 1136, [2007] Bus LR 1182, 112 ConLR 1, (2007) Times, 8 May, 2007 SC (HL) 116, 2007 SCLR 429, 151 Sol Jo LB 571, [2007] All ER (D) 226 (Apr), 2007 Scot (D) 9/2
33 Section 14(3)

6   The adjudication process

Section 6 consists of 18 very short subsections. It, along with the Code of Practice, comprises the law on adjudication. Section 7 deals with the entitlement to suspend work for failure to comply with an adjudicator's decision; section 8 provides for a panel of adjudicators being appointed by the Minister; and section 9 entitles the Minister to publish a Code of Practice governing the conduct of adjudications under section 6. These provisions comprise the skeletal framework for adjudication.

Subsection 6(1): Entitlement to adjudication

(1) *A party to a construction contract has the right to refer for adjudication in accordance with this section any dispute relating to payment arising under the construction contract (in this Act referred to as a 'payment dispute').*

Adjudication is optional

The entitlement to adjudicate is just that; it is an entitlement and not an obligation. If the claimant believes that it is better served by availing of other means of dispute resolution as provided for in the contract, or through litigation, there is nothing in the Irish Act to prevent this (*Cubic Building & Interiors Limited v Richardson Roofing (Industrial) Limited*).[1] Whether or not a contractual provision for adjudication is mandatory or optional depends upon the wording of the provision. In the UK adjudication is often mandatory by reason of the wording of the contract.[2] This is not seen as opting out of the statute.

'Arising under'

The House of Lords in the *Fiona Trust* case[3] considered the words 'arising under' in connection with an arbitration clause and held that they meant the same as 'in connection with', 'in relation to' and 'out of', and should be given purposeful interpretation. In upholding the validity of the arbitration clause in order to determine any dispute arising out of the contract, the House of Lords started with the assumption that the parties, as rational businessmen, were likely to have intended any dispute arising out of the relationship into which they had entered or purported to enter to be decided by the same tribunal.

The High Court in New Zealand took a different view in *Jian Hua Property Limited v Freemont Design and Construction Limited.*[4] In that case, a contract had to be abandoned because of defects in the planning. The contractor sought to include claims for loss of profits arising from the fact that it could not complete the works. This was held by the court not to be a claim arising 'under the contract' and therefore to be outside the scope of statutory adjudication.

In *Redhill Development (NZ) Limited v Green*[5] Lang J of the High Court of Auckland stated:

> I deal first with the submission that the phrase 'under the contract' should be interpreted narrowly. I agree that the use of this phrase suggests that Parliament intended to restrict the range of disputes that adjudicators could determine under the Act. Disputes may arise out of, or in relation to, a construction contract in numerous ways. It would not be appropriate, however, for many of them to be determined by an adjudicator. Claims for misrepresentation and under the Fair Trading Act 1986 are good examples of this.

The New Zealand and the UK legislation allow any dispute to be referred to adjudication irrespective of whether it relates to payment. It is to be noted, however, that even in those countries the dispute must arise 'under the contract'. Claims in respect of extensions of time, the valuation of defects and the valuation of variations would all arise under the contract as the standard forms of contract invariably provide for these. There are many circumstances, however, that may give rise to claims that are not strictly speaking 'under the contract'. Whether or not the Irish court would follow the precedent set by the House of Lords in *Fiona Trust* in interpreting the intention of the parties when availing of words such as "*disputes arising under the contract*"[6] remains to be seen. A distinction has to be drawn however between the presumed commercial intent of the parties to a contract on the one hand, and a statutory provision on the other. The Irish courts are likely to take a more restrictive approach to the latter.

'Relating to payment'

The entitlement to adjudication relates only to disputes 'relating to payment'. The corresponding section in the UK Act of 1996 relates simply to 'disputes'. This begs the question as to the extent to which the legislature intended to confine the nature of the dispute. On a narrow interpretation, the term might be defined by reference only to disputes arising out of Section 4 of the Irish Act. The general view, however, is that the term 'relating to payment' is by its nature a very wide expression and that it will be interpreted liberally. Most disputes relating to construction contracts do relate to payment in one form or another. A claim for extension of time for instance is almost invariably accompanied by a claim for compensation.

Presumably, such claims will be regarded as claims relating to payment. Issues such as the powers of the certifier, when those powers do not relate to payment, would obviously be excluded. No doubt a myriad of borderline issues will arise.

One such borderline issue might relate to disputes about whether a particular form of contract has been incorporated into the agreement between the parties. Parties often describe the form of contract they intend to incorporate in loose and inaccurate terms. For instance, a statement that 'the CIF Form of Sub-contract will apply' could relate to the sub-contract issued by the Construction Industry Federation and the Sub-Contractors and Specialists Association, fifth edition, October 1989, or the Specialist Contract Document issued by the Construction Industry Federation following agreement between the Master Builders and Contractors Association and Alliance of Specialist Contractors Association 1999. It is clear that this issue of itself would not be covered by the Irish Act. One would assume, however, that if it is necessary to resolve this issue in order for the adjudicator to decide upon a dispute relating to payment, the adjudicator would have jurisdiction to do so. This seems reasonably clear. It would normally be necessary to establish the terms of the contract to calculate the sum due.

On the other hand, disputes often also arise as to the identity of the contracting parties. It would not be necessary for the adjudicator to decide that issue for the purpose of calculating the payment. If the adjudicator is appointed in relation to a dispute between two named parties, and one of the parties is alleging that they are not a party to the contract, this would go to the jurisdiction of the adjudicator. There is no provision in the legislation which entitles adjudicators to bind the parties by their decisions on jurisdiction. The question therefore that arises is whether an adjudicator, in deciding a dispute relating to payment, is by virtue of that dispute entitled to decide the identity of the party to whom payment is to be made. It is likely that a liberal approach to the legislation by the courts would confirm that in these circumstances an adjudicator would be so entitled.

The legislation applicable in the Australian States and Territories provides invariably for the making of a progress payment claim and for the referral to adjudication of any dispute arising from the payment claim. For instance, the Queensland Act provides at subsection 17(1) for the service of a payment claim by a person: 'who claims to be entitled to a progress payment'. One might consider a dispute arising from a: 'progress payment claim' as being, if anything, narrower in its application to: 'any dispute relating to payment' being the expression used in subsection 6(1) of the Irish Act. Yet in Queensland, adjudicators have always been seen as having the jurisdiction to deal with disputes relating to latent defects and time-related claims. Without making any amendment to the relevant provisions of the 2004 Act, and in particular without amending section 17(1), the 2014 Queensland amending legislation has labelled such disputes as complex payment claims. Therefore, such claims already fell within the jurisdiction of an adjudicator under the terms of the earlier legislation. Assuming the Irish courts take a similar view, it must follow that: 'any dispute relating to payment' will be interpreted as including claims for extension of time, and, indeed, claims of any nature provided a resolution is necessary to ascertain a financial consequence.

Despite the fact that the entitlement to adjudication generally in Australia is confined to payment disputes, and the legislation in the UK is not, the distinction has not given rise in practical terms to any difference in the impact of adjudication in these different jurisdictions. It is unlikely, therefore, that the fact that the Irish Act is confined to disputes relating to payment will in practical terms have any restrictive effect on the extent to which adjudication will be available for the resolution of disputes.

If it was intended by the Irish legislature to confine the nature of the dispute that could be referred to adjudication strictly to those arising out of progress payment claims, the precedents were there for the legislature to follow in the form of the Australian East Coast Model and the Singapore Act. The Singapore Act at section 10 entitles a claimant to serve a payment claim in respect of a progress payment. Section 11 requires the respondent (defined as a person who may be liable to make a progress payment) to make a response, and section 12 entitles the claimant to refer the dispute to adjudication unless the respondent accepts and pays on foot of the claim. It is clear, therefore, that adjudication is only available to the executing party and that it only relates to a progress payment claim. This is in stark contrast with the entitlement under the Irish Act to refer any dispute relating to payment for adjudication.

There is in fact a disconnect between the payment provisions and the adjudication entitlement under the Irish Act. The dispute the subject matter of the adjudication need not necessarily arise from a progress payment claim. This is not unusual. The same is the case with the New Zealand Act and with the UK legislation, where a party to the contract may refer any dispute, whether related to a payment claim or not, to adjudication.

One drawback to confining adjudication to disputes relating to payment is that claimants may not be able to subdivide a dispute so that issues of liability are dealt with in the first instance and quantum thereafter. It is not unusual in the UK for a claimant to seek adjudication purely on issues of liability in the first instance. If it is successful, it will then present a separate claim in respect of the quantum, which may well be heard before a different adjudicator. This process makes a lot of practical sense in terms of time, efficiency and cost.

What is a 'dispute'?

The term 'dispute' is not defined. One of the leading cases on this issue is that of *Amec Civil Engineering Limited v Secretary of State for Transport.*[7] Jackson J provided the following guidance, which has been favoured by the Court of Appeal of England and Wales in subsequent cases:[8]

1   The word 'dispute' which occurs in many arbitration clauses and also in s 108 of the Housing Grants Act should be given its normal meaning. It does not have some special or unusual meaning conferred upon it by lawyers.
2   Despite the simple meaning of the word 'dispute', there has been much litigation over the years as to whether or not disputes existed in particular

situations. This litigation has not generated any hard-edged legal rules as to what is or is not a dispute. However, the accumulating judicial decisions have produced helpful guidance.

3   The mere fact that one party (whom I shall call 'the claimant') notifies the other party (whom I shall call 'the respondent') of a claim does not automatically and immediately give rise to a dispute. It is clear, both as a matter of language and from judicial decisions, that a dispute does not arise unless and until it emerges that the claim is not admitted.

4   The circumstances from which it may emerge that a claim is not admitted are protean. For example, there may be an express rejection of the claim. There may be discussions between the parties from which objectively it is to be inferred that the claim is not admitted. The respondent may prevaricate, thus giving rise to the inference that he does not admit the claim. The respondent may simply remain silent for a period of time, thus giving rise to the same inference.

5   The period of time for which a respondent may remain silent before a dispute is to be inferred depends heavily upon the facts of the case and the contractual structure. Where the gist of the claim is well known and it is obviously controversial, a very short period of silence may suffice to give rise to this inference. Where the claim is notified to some agent of the respondent who has a legal duty to consider the claim independently and then give a considered response, a longer period of time may be required before it can be inferred that mere silence gives rise to a dispute.

6   If the claimant imposes upon the respondent a deadline for responding to the claim, that deadline does not have the automatic effect of curtailing what would otherwise be a reasonable time for responding. On the other hand, a stated deadline and the reasons for its imposition may be relevant factors when the court comes to consider what is a reasonable time for responding.

7   If the claim as presented by the claimant is so nebulous and ill-defined that the respondent cannot sensibly respond to it, neither silence by the respondent nor even an express non-admission is likely to give rise to a dispute for the purposes of arbitration or adjudication.

The courts in the UK have been very slow in practice to set aside an adjudicator's decision simply because a dispute had not crystallised. The courts have taken a very pragmatic approach – provided the claim has been stated, a dispute will have arisen if the claim is not admitted within a relatively short period. However, the claim does have to be stated. In *Allied P&L Limited v Pardigm Primary Housing Group Limited*[9] Akenhead J came to the conclusion that although the claimant had disputed the respondent's termination of the contract, the claimant had not prior to the adjudication put forward any claims for compensation arising out of the termination. He found that there was no dispute in relation to these financial claims. However, the adjudicator's decision was upheld on the financial claims because the respondent had responded to those claims and taken part in the adjudication of them without reserving its position. However in *Beck Interiors Ltd v*

UK Flooring Contractors Ltd[10] the Court found that the adjudicator did not have jurisdiction in respect of a claim that had not crystallised into a dispute because sufficient time had not passed between the claim being made and the notice of adjudication being issued to infer a rejection of the claim.

Although subsection 6(1) is stated in terms of a single dispute, subsection 6(9) does empower the adjudicator to deal at the same time with several payment disputes. It is well settled in the UK that a dispute as to the amount due on a final account, which might rest on numerous differences ranging from extension of time claims to claims for variations and the cost of inflation, are all to be considered as a dispute for the purpose of the UK Act of 1996.[11] This is in keeping with the general approach by the courts in that jurisdiction to upholding the entitlement to adjudication and therefore a swift determination, irrespective of the size or complexity of the dispute, and notwithstanding the difficulties that such disputes give rise to on the part of the respondent and the adjudicator.

It is sometimes argued that the fact the parties are in negotiation is indicative that a dispute has not yet crystallised. The Court of Appeal in the case of *Collins (Contractors) Limited v Baltic Key Management 1994 Ltd*[12] came to the opposite conclusion, the fact that the parties are in negotiation is indicative that a dispute exists.

It would appear that the very fact that an application is made for payment under a contract and that a response is not received to the application within the time required by the contract (or the Schedule to the Irish Act if appropriate), will give rise to a dispute for the purpose of the Irish Act. In *Geoffrey Osborne Limited v Atkins Rail Limited*[13] Edwards-Stuart J at paragraph 50 found:

> The ground on which GOL claimed to be entitled to refer the dispute to adjudication was ARL's failure to issue an Interim Certificate within 14 days of Payment Application No 36. GOL was entitled to have its claims assessed within 14 days and this had not happened. It was therefore entitled to refer the matters covered by its application for payment to adjudication so that they could be assessed by an adjudicator. It is not now contended that ARL's failure to issue a certificate within the required 14 days did not give rise to a dispute (if it had been, I would have unhesitatingly held that it did).

Standard contracts usually provide for the contract administrator having a period of time to consider the contractor's claim prior to making a determination in respect of it. Given that the parties through their contract have agreed to this process it cannot be said that a dispute has arisen until after that period of time has expired.

Ambush

To avoid adjudication being availed of to ambush respondents, initially the courts in England interpreted the word 'dispute' very narrowly. In *Edmund Nuttall v RG Carter*[14] at paragraph 36, Seymour J stated:

The whole concept underlying adjudication is that the parties to an adjudication should first themselves have attempted to resolve their differences by open exchange of views and, if they are unable to, they should submit to an independent third party for decision, the facts and arguments which they have previously rehearsed among themselves. If adjudication does not work in that way there is the risk of premature and unnecessary adjudications . . . There is also a risk that a party to an adjudication might be ambushed by new arguments and assessments which have not featured in the 'dispute' up to that point.

A number of decisions called into question the validity of the *Nuttall* judgment and, in light of subsequent judgments in cases such as that of *Amec* referred to above, *Cantillon v Urvasco*[15] and *Bovis Lend Lease v The Trustees of the London Clinic*,[16] it must be assumed that the *Nuttall* case would not be followed in Ireland. The Irish courts are likely to take the more liberal approach adopted in these cases, whereby a dispute is deemed to have arisen if it can be said that the parties, in normal parlance, are in dispute. The courts have taken the view that a respondent is not to be protected against ambush by an artificially rigid interpretation of the word 'dispute'. Instead, the respondent will be protected by the court's insistence that the respondent has a proper opportunity to meet the case made against it. If a respondent is ambushed and cannot quite genuinely meet the case against it within the time allowed by statute, either the adjudicator, or the court may well come to their rescue – the adjudicator by not making a decision against a party who has not had an opportunity to properly challenge the claim, and the courts by refusing to enforce an adjudicator's decision where the tenets of natural justice have not been met.

The legislation in most countries, including Ireland, requires adjudicators to make their decisions within the statutory period. It also requires adjudicators, either expressly or impliedly to act impartially. The law requires adjudicators to apply fair procedures, but the legislation does not give any assistance to adjudicators as to how they are to resolve the conundrum that arises when it is clear to the adjudicator upon receipt of the adjudication referral that it will be impossible, because of the complexity of it, to deal with the matter fairly within the time prescribed.[17] The guidance notes for adjudicators published in the UK jointly by the Adjudication Society and the Chartered Institute of Arbitrators state:

> An adjudicator should decide whether or not he is capable of arriving at a fair conclusion within the limited period available to him. If he is not, then he should either decline the reference, or (as is frequently done in practice) accept the reference subject to receiving what appears to be an appropriate extension of time from the parties (both parties if the extension of time required is more than fourteen days) at the beginning of the process.[18]

Obviously, if adjudicators believe that they would be unable, because of other commitments or limited capacity, to reach a decision within the time prescribed, they should decline the reference. A more difficult issue arises if the adjudicator is of the view that it would be impossible for any adjudicator to deal fairly with the matter within the time prescribed and the claimant will simply not allow a sufficient extension. The legislation is drafted in such a way that a claimant who is not prepared to consent to an extension of time is entitled to a decision within 28 days. Is it appropriate for adjudicators to set that statutory entitlement to nought by refusing to accept the reference where they are of the belief that justice cannot be done within the prescribed period? The paragraphs quoted below on page 70 from the judgment of Chadwick LJ in *Carillion Construction Limited v Devonport Royal Dockyard Limited*[19] might suggest that the adjudicator need only be fair to the extent that the legislation permits, and that on occasion there must be an acceptance that rough justice will prevail. On the other hand, the courts have also endorsed the principle that an adjudicator should ensure at the outset that the dispute is capable of being dealt with fairly within the time provided.[20] The fact that the courts in the UK have never refused to enforce an adjudicator's decision on the basis that the dispute was too complex for the adjudicator to do justice to the parties in 18 years, would suggest that, at least in that jurisdiction, this is not a ground with which the courts are likely to find favour.

A refusal, upon receipt of the referral notice, by an adjudicator not to proceed with the adjudication unless the time for the adjudicator's decision is extended could in fact overcome the potential injustice to the respondent if the claimant refused to allow the extension. In those circumstances a new adjudicator would have to be appointed but the respondent would already have the referral notice and would therefore have extra time to consider its response in the period taken to have a new appointment made. That of course does not get over the problem that the adjudicator's difficulty may not relate to the time available to the respondent but to the time available to the adjudicator to deal fairly with the dispute.

The fact that the courts in the UK have not faulted adjudicators for not refusing to proceed with adjudications in the knowledge that it was unlikely that they would be able to do full justice to the parties in view of the complexity of the dispute and the time available to give their decision, would tacitly suggest that it is pardonable for adjudicators to proceed in these circumstances. If that assumption is correct, the Northern Territory of Australia Act constitutes an exception. Section 33(1)(iv) requires the adjudicator to dismiss the application without making a determination of its merits if the adjudicator is: 'satisfied that it is not possible to fairly make a determination:

a because of the complexity of the matter; or
b because the prescribed time or any extension of it is not sufficient for another reason'.

Subsection 6(2): Notice of Intention

(2) *The party may exercise the right by serving on the other person who is party to the construction contract at any time notice of intention to refer the payment dispute for adjudication.*

Subsection 10(1) of the Irish Act provides that the construction contract may set out the manner in which notices are to be delivered. If there is no such agreement, the notice under subsection 10(2) is to be delivered 'by post or by any other effective means'.

Subsection 6(2) could not be wider. The notice can be served at any time. However, it must set out the nature of the dispute and the reliefs sought. The full requirements of the Notice of Intention are set out in the Code of Practice at paragraph 5.

When the UK Act of 1996 was debated in Parliament prior to its enactment, there was a school of thought that the right to refer a dispute to adjudication should only arise up to the date of practical completion of the contract. The rationale was that the vast majority of disputes prior to practical completion would be relatively straightforward cash flow issues. More complex issues, such as the entitlement to extension of time and the compensation payable for extension of time, were more likely to crystallise after practical completion and were better dealt with through other forms of dispute resolution. However, the inclusion of the words 'at any time' in the Irish Act and in the UK Act of 1996 make it clear that the right to adjudication continues after practical completion.

Although in keeping with the UK legislation, the right to seek adjudication at any time is not a commonplace entitlement. For instance, the Queensland Act (prior to the 2014 amendment) required that the application must be made for adjudication within ten business days of receiving the payment response ('schedule'). It is also common practice in Australia to confine the entitlement to a relatively short period following completion of the works.

In the eight jurisdictions of Australia providing for adjudication, there is considerable divergence as to the latest date a claimant should be permitted to make a payment claim, i.e. a claim in respect of which adjudication may be sought. The periods range from 28 days following the last day on which construction works were carried out, to 12 months from that date, the latter generally to coincide with the ending of the defects liability period. Victoria provides for a period of three months and South Australia for a period of six months. The Wallace Report gave careful consideration to this issue and recommended for the State of Queensland that in respect of interim progress payments, the latest date should be 6 months from the date construction works were last carried out (or goods and services supplied), but in the case of a final account payment the latest date should be 28 days after the end of the defects liability period under the construction contract.[21] These recommendations have been incorporated into the amending legislation. The pressure to provide a shorter timeframe arises from a perception that a lengthy period of time encourages ambush, i.e. the claimant can spend twelve months following practical completion putting together a very comprehensive claim to which the respondent must respond in a very short period of time.

The Notice of Intention is of the utmost importance. It is from the notice that adjudicators derive their jurisdiction. Adjudicators must be careful to ensure that they do not exceed their jurisdiction by deciding some element of a dispute between the parties that is not covered by the Notice of Intention. A difficult question for adjudicators is often whether issues raised in a defence in fact constitute new issues outside of the adjudicator's jurisdiction. If the defence includes a cross claim that has already been set up by way of a response to a claim made under section 4 of the Irish Act and the details under subsection 4(4) have been provided, the adjudicator's jurisdiction will cover such cross claims. Where, however, the cross claims have not been included in such a response, they will generally be outside the adjudicator's jurisdiction. The proper course for the respondent in such circumstances is to pursue a claim in relation to those cross claims independently of the claimant's Notice of Intention. In the meantime, the respondent will be obliged to pay any money found by the adjudicator to be due on foot of the claimant's Notice of Intention.

In the English case of *Mecright Limited v TA Morris Developments Limited*[22] the claimant (TA Morris) contended that it had lawfully terminated the sub-contract and was entitled to damages. The respondent alleged that the claimant had in fact repudiated the sub-contract and sought damages in the adjudication for that reason. The adjudicator found that the claimant had repudiated the contract, which decision was within the adjudicator's jurisdiction as an invalid termination would be a repudiation. However, without jurisdiction, the adjudicator determined the monetary compensation to which the respondent was entitled on foot of the repudiation. Whereas *TA Morris* had claimed in the Notice of Adjudication for a declaration that it had lawfully terminated the sub-contract and for damages resulting from the termination, there was no claim before the adjudicator on the part of *Mecright Limited* for damages on a proper construction of the Notice of Adjudication. The court therefore refused to enforce payment on foot of the adjudicator's decision.

The question arises as to whether an adjudicator appointed in relation to a second dispute is bound by the decision of an adjudicator in an earlier dispute. In arbitration, of course, a subsequent arbitrator is bound by the decision of the earlier arbitrator. The same is the case in adjudication, on matters of principle. Therefore, if an adjudicator finds that a contractor is not entitled to an extension of time based on a specific set of facts, the decision is binding upon any subsequent adjudicator. However, if a subsequent adjudicator is faced with a new claim for extension of time based on entirely different facts, albeit coinciding or overlapping with the earlier claim, they may be entitled to find that the contractor is so entitled (*Emcor Drake & Scull Limited v Costain Reconstruction Limited*).[23] This matter is discussed in greater detail below in the context of subsection 6(10).

The question also arises as to whether a party to a construction contract is entitled to refer more than one dispute, or indeed numerous disputes, to adjudication at the same time. Mr Justice Ramsey of the Technology and Construction Court found in *Wilmott Dixon Housing Limited (formerly Inspace Partnerships Limited) v Newlon Housing Trust*[24] that having regard to the words 'at any time', a party is entitled to refer a number of disputes to adjudication at the same time. In an Irish context, one must bear in mind the content of subsection 6(9), which entitles the adjudicator to: 'deal at the same time with several payment disputes arising under the same construction contract or related construction contracts'.

In the English case of *Connex South Eastern Limited*[25] the term 'at any time' was given literal meaning, and it was not an abuse of process to serve notice 15 months after termination, even if it was no longer possible to have a quick and cheap temporary decision. In the Court of Appeal, Lord Justice Dyson, delivering the judgment of the court, stated:

> The phrase 'at any time' means exactly what it says. It would have been possible to restrict the time within which an adjudication could be commenced, say, to a period by reference to the date when work was completed or the contract terminated. But this was not done. It is clear from Hansard that the question of the time for referring a dispute to adjudication was carefully considered, and that it was decided not to provide any time limit for the reasons given by Lord Lucas. Those reasons were entirely rational.
>
> There is, therefore, no time limit. There may be circumstances as a result of which a party loses the right to refer a dispute to adjudication: the right may have been waived or the subject of an estoppel. However, subject to considerations of this kind, there is nothing to prevent a party from referring a dispute to adjudication at any time, even after the expiry of the relevant limitation period. Similarly, there is nothing to stop a party from issuing court proceedings after the expiry of the relevant limitation period. Just as a party who takes that course in court proceedings runs the risk that, if the limitation defence is pleaded, the claim will fail (and indeed may be struck out), so a party who takes that course in an adjudication runs the risk that, if the limitation defence is taken, the adjudicator will make an award in favour of the respondent.

In relation to the argument that Parliament intended adjudication to be quick and inexpensive, Lord Justice Dyson stated:

> I can accept that Parliament intended adjudication to be quick and (relatively) cheap, although it may not have been entirely successful in bringing this about. But that says nothing about when the quick and (relatively) cheap adjudication may be commenced. There is no link between the speed and expense of an adjudication and the time when it starts. An adjudication started before practical completion may be complex, slow and expensive. Conversely, an adjudication started long after practical completion may be simple, quick and cheap. Nor do I understand why an adjudication conducted long after practical completion cannot on that account result in a decision which has provisional or temporary effect only.

The legislation applicable in the UK and in Ireland therefore permits the commencement of adjudication proceedings at any time, irrespective of the period of time that may have elapsed since the payment claim notice was given and/ or the contract works completed. This is contrary to the approach taken in the Australian states, where the entitlement to statutory adjudication only arises if notice is given promptly after the rejection of the claim or, in the case of a final account, reasonably promptly after the completion of the works or the completion of the defects' liability period, depending on the state concerned.

The fact that arbitration proceedings are already in being in relation to the same dispute does not prevent the issue of a Notice of Adjudication. In *Cubitt Building and Interiors Ltd v Richardson Roofing (Industrial) Ltd*[26] a requested stay on arbitration pending adjudication was refused. In refusing to restrain the arbitration proceedings, Akenhead J of the TCC noted:

> Of course, it is open to any party to apply for relief to the requisite tribunal to enable it to exercise its right to adjudicate. I do not accept however that there must be a stay of any legitimately constituted proceedings, whether in arbitration or in court proceedings, where there is merely a discretionary right to adjudicate as opposed to a binding pre-conditional adjudication requirement. I suspect that what the learned judge and author really intended was that the proceedings in question, in terms of timetable and the like, should not be so conducted as to prevent a party from pursuing its contractual or statutory right to adjudicate. Thus, it may be appropriate in certain circumstances to build into the timetable in court or arbitration proceedings a 28-day period to enable one party to adjudicate if, for any good reason, it cannot sensibly pursue adjudication at the same time as its court or arbitration proceedings. Thus, having regard to the Overriding Objective, if the court believes, following representations, that there is a measurably good prospect that adjudication will finally resolve the disputes or some of them the court may well build into its timetable for trial some time to enable a party to adjudicate. That however is different from a stay. A party who has started court or arbitration proceedings is entitled to have those proceedings resolved as reasonably expeditiously as the court can achieve and justice demands; it should not be forced to have those proceedings delayed or stayed by it itself being forced to adjudicate when it does not want to exercise its right to do so.

In most jurisdictions, other than those of the UK and Ireland, the legislation expressly provides for the entitlement to pursue adjudication and other remedies under the contract, or at law, at the same time. For example the New Zealand Act at section 26 states:

1 To avoid doubt, nothing in this Part prevents the parties to a construction contract from submitting a dispute to another dispute resolution procedure (for example to a court or tribunal, or to mediation), whether or not the proceedings for the other dispute resolution procedure take place concurrently with an adjudication.

2 If a party to a construction contract submits a dispute to another dispute resolution procedure while the dispute is the subject of an adjudication, the submission to that other dispute resolution procedure does not –

    (a) Bring to an end the adjudication proceedings; or

    (b) Otherwise affect the adjudication.

The various bodies responsible for publishing standard forms of contract are already considering changes that might arise by reason of the Irish Act. One of the changes to be considered is whether participation in adjudication is to be considered a necessary prerequisite to commencing arbitration proceedings or litigation. The effect of such a provision would circumvent the optional nature of adjudication as provided for by the Irish Act. It would also enable a party to obtain a stay on arbitration or court proceedings if the claimant failed to proceed to adjudication in advance of arbitration or litigation. It is submitted that a contractual obligation to commence adjudication in advance of other forms of dispute resolution would be enforced by the Irish courts, but that any provision, which sought to prevent adjudication taking place subsequent to any other procedure being commenced, would not be enforceable. Contractual provisions, which extend or enhance the application of the legislation, are likely to be found acceptable; those that seek to alter the applicability of the Act, however, will not.

As indicated, the entitlement to refer a dispute at any time to adjudication includes a right to refer the dispute to adjudication when it is already the subject matter of proceedings in arbitration or before the courts (*Herschel Engineering Limited v Breen Property Limited*)[27]. This makes perfect sense. It may become apparent that arbitration proceedings already commenced will be protracted and that a hearing may not be possible before some distant date or, in the case of court proceedings, a High Court decision may be appealed to the Supreme Court and the appeal may not be heard for a number of years. In the meantime, for the sake of cash flow, the claimant is entitled to have the matter processed through adjudication. It would appear, therefore, that the principles applying from such cases as *Henderson v Henderson*[28] do not necessarily apply to adjudication, at least in this context. That is not to say that issue estoppel does not apply to adjudication. A party is not entitled to re-agitate the same dispute on a different legal basis in adjudication any more than it would be entitled to do so in arbitration or litigation. In *Melbourne Authority v Anshun PTY Limited*[29] McDougal J of the New South Wales Court of Appeal found that the claimant was not entitled to seek to have claims already dealt with in adjudication heard by a second adjudicator simply on the basis that the legal grounds to justify the claims were different. The court found that, except in special circumstances, a party was obliged to put the entire of its case and all of its grounds before the adjudicator and that, in effect, the decision of the adjudicator would be as binding as if all the grounds that could have been relied upon were relied upon.

However, decisions in principle that are binding upon successive adjudicators inter se are not, of course, binding on arbitrators or the courts, and it is for this reason that it is possible to have arbitration and adjudication proceedings running concurrently. The adjudicator's decision will have effect temporarily until the arbitrator's award is made.

Subsection 6(3) and (4): Appointment of the adjudicator

(3) *The parties may, within 5 days beginning with the day on which notice under subsection (2) is served, agree to appoint an adjudicator of their own choice or from the panel appointed by the Minister under section 8.*

(4) *Failing agreement between the parties under subsection (3), the adjudicator shall be appointed by the chair of the panel selected by the Minister under section 8.*

It would appear that neither party can seek to have an adjudicator appointed
from the panel until the period of five days expires (paragraph 14 of the Code of
Practice). There is, however, no mandatory requirement for either party to nominate other potential adjudicators. No doubt parties will seek to agree upon an
adjudicator before seeking an appointment by the Minister, but this is optional.
If the nature of the dispute being referred to adjudication is unclear, the respondent should seek information immediately upon receipt of the Notice of Intention
rather than leave it until receipt of the Referral Notice.

In the UK appointments are made, in default of agreement, by approved professional institutions. This system has worked well. The impartiality of the appointing bodies and the standard of adjudicators' decisions have generated a sense of
confidence in the industry. This is not the case in Australia. The fact that appointments are made there in default by profit-making nominating bodies chosen by the
claimant has led to perceived bias on the part of adjudicators. The Wallace Report
identified a mistrust of the adjudication process amongst respondents arising from
the fact that the claimant was in a position to select the nominating body to make
the appointment. Claimants were therefore likely to gravitate towards nominating
bodies whose adjudicators were considered to be claimant friendly. At page 160
Mr Wallace observed:

> In my view, the nexus between the interests of claimants and their representa-
> tives must be broken from those that appoint adjudicators and of course the
> adjudicators themselves. Putting aside all of the untested allegations made to
> the Review and applying the test adopted in Johnson v Johnson[30] it is my view
> that a fair minded lay observer might reasonably apprehend that an adjudica-
> tor under the current appointment process might not bring an impartial and
> unprejudiced mind to the resolution of the questions that he or she is called
> upon to decide.
>
> While I respect that in a perfect world, the parties should be able to agree
> upon an appropriate (appointing body) or adjudicator, I am not convinced
> that such a process would not be open to abuse. Given that in most cases a
> contracting party will be responsible for the drafting of a building contract,
> I am concerned that respondents would simply name 'respondent friendly'
> (appointing bodies) or 'respondent friendly adjudicators'. The mischief that
> many respondents have so bitterly complained of would simply be replicated
> by such a process, but this time in their own favour.

As with the States of Victoria and New South Wales, therefore, the Queensland
reforming legislation has moved away from appointments being made by approved
nominating bodies and towards a central registrar administering and monitoring
not only the appointment of adjudicators to disputes but also the performance and
continued registration of such adjudicators on a central panel.

The fact that adjudication may be commenced at any time allows a claimant to organise events so that the respondent is obliged to respond to the claim over the Christmas holiday period, which is the classical ambush. In Queensland, prior to the amending legislation of 2014, time constraints were measured in terms of business days and that term was defined so as to exclude the period from 25 December to 1 January inclusive. The 2014 Act increases this exclusion by adding the three days before Christmas and the three days subsequent to 1 January. A similar exclusion, it is suggested, would be highly appropriate in Ireland.

In recommending these additional days, the Wallace Report also considered excluding some days around Easter, but decided against this. The main construction holiday period in Australia is between 25 December and 1 January. Whereas the main holiday period in the construction industry is the last fortnight of July in Ireland, management do not have any particular difficulty in dealing with claims in that period. As with Australia, it is the Christmas holidays that are particularly difficult.

It would appear that an agreement between the parties in a contract to have an adjudicator appointed, in default of agreement, by an institution, such as Engineers Ireland, would be invalid. It would appear that if the parties cannot agree upon an adjudicator, the only body authorised to make the appointment is the chair of the panel. To agree otherwise would arguably fall foul of subsection 2(5). On the other hand, it may be argued that because the Irish Act does not expressly say that any agreement of this nature would be invalid, such an agreement would be enforceable. A comparison might be drawn between such a provision and the entitlement (assuming the courts decide there is an entitlement) of the parties to include in their contract a mandatory requirement for adjudication as a prerequisite to arbitration or litigation. Such a mandatory requirement removes the discretionary entitlement to proceed, or not to proceed, by way of adjudication conferred by subsection 6(1). Given, however, that there is no list of approved institutions in Ireland or any means under the Act for identifying such a list, it is thought unlikely the courts would permit of the parties agreeing in their contract on a particular institution to make the appointment. This would give rise to the very dangers sought to be avoided through the amending legislation in Queensland. Whilst this matter seemed quite clear prior to the publication of the Code of Practice, some doubt has been raised by the inclusion in paragraph 8 of the reference to a person being named in the construction contract. The matter is discussed below by reference to paragraph 8 in Chapter 12.

Likewise, it would appear that any agreement between the parties as to the appointment of an adjudicator would be unenforceable if not made within the five-day period referred to in subsection 6(3). If, therefore, the parties had agreed in their contract that a particular person would be the adjudicator for any disputes that may arise, subsection 6(3) would appear to afford an opportunity to renege upon that agreement. In view of the findings of the Wallace Report as quoted above, it would be unfortunate if the courts took the view that parties are entitled to agree through their contract on the person who is to adjudicate their disputes and that such a provision would be binding. Given, however, that there is no express provision in the Irish Act preventing this, the issue is open to debate. The balance of that debate may be shifted to some degree in favour of the parties being entitled to appoint an adjudicator through their contract by reason of such an entitlement being endorsed by the Code of Practice at paragraph 8.

Subsection 6(5): Referral of the dispute to the adjudicator

(5) *The party by whom the notice under subsection (2) was served—*

(a) *shall refer the payment dispute to the adjudicator within 7 days beginning with the day on which the appointment is made, and*

(b) *shall at the same time provide a copy of the referral and all accompanying documents to the person who is party to the construction contract.*

The necessary content of the referral is set out in the Code of Practice at paragraph 22. The Irish Act does not set out the consequences of a failure on the part of the referring party to refer the dispute to the adjudicator within seven days. Is the other party entitled to disregard the appointment if the referral does not occur within that time? It would appear that it would be so entitled (*PT Building Services Limited v ROK Build Limited*[31] and *Hart Investments v Fidler*[32]). In the former case, Ramsay J opined at paragraph 54:

> In my judgment, the central purpose of the Scheme is to incorporate those fundamental provisions which, when absent, lead to the Scheme being imposed as an implied term. The provision in paragraph 7(1) of Part 1 of the Scheme, which was considered in Hart v Fidler, is derived from Section 108(2)(b) of the Act. That, it seems to me, makes paragraph 7(1) of the Scheme one of the fundamental provisions in the process of adjudication.

The central issue in the *PT Building Services* case was not whether failure to serve the Referral Notice within the required statutory period was fatal to the process, but whether failure to provide the accompanying documents at the same time would also be fatal to the process. The court in the paragraph quoted above went on to distinguish a breach of paragraph 7(1) from a breach of paragraph 7(2) on the basis that paragraph 7(2) is merely 'an associated procedural requirement'. In that respect, the court found:

> 55. I consider that it is undesirable that every breach of the terms of the Scheme, no matter how trivial, should be seized upon to impeach the process of adjudication. To do so would increase the tendency of parties to take a fine tooth-comb to every aspect of the adjudication in the hope of finding some breach of the Scheme on which to impeach an otherwise valid adjudication decision. I do not consider that that was either intended or the natural effect of a failure to comply with the Scheme. There may, of course, be cases where the documents included with the

referral notice are so deficient that it effects [sic] the validity of the adjudication process. However, I do not consider that a failure to include the relevant construction contract until a day later can do so or does so on the facts of this case. Nor do I consider that a failure to include the construction contract can be said to amount to such a serious breach of the rules of natural justice that the decision should not be enforced. There is nothing obviously unfair in the documents relied on in relation to the construction contract being received by the adjudicator later than the referral notice: see Carillion v. Devonport [2006] BLR 15 at paragraph 85.

Paragraphs 7(1) and 7(2) of Part 1 of the Scheme are not identical to, but are the equivalent of, subsections 6(5)(a) and (b). It is likely, therefore, that the Irish courts would take the view that if the accompanying documents are provided shortly after the referral, but not at the same time, this will not be fatal to the process.

There is, however, a difference in wording between subsection 6(5)(b) of the Irish Act and the equivalent provision in the UK Act of 1996. Section 108(2)(b) of the UK Act of 1996 requires the dispute to be referred to the adjudicator within seven days of the Notice of Adjudication. The Irish Act requires that the dispute be referred to the adjudicator within seven days of the appointment. The Irish legislation imposes a pause on the process to enable the parties to agree upon an adjudicator or, in default of agreement, to enable an appointment to be made from the panel of adjudicators to be appointed by the Minister. Under paragraph 18 of the Code of Practice, the chair of the panel is to make the appointment *"normally"* within seven days of the request being made, but there is no time limit imposed by the Act or by the Code of Practice for making such a request.

If the respondent, despite the referral being made outside the seven days, fully participates in the adjudication, it will have lost the right to object to any determination on this basis (*KNN Coburn LLP v GD City Holdings Limited*)[33] (see below at page 72).

Response to the referral

The Irish Act does not expressly provide for the respondent delivering a response to the referral. In most jurisdictions there is a specific requirement on the respondent to deliver a response to the referral to adjudication within a specific time. In these jurisdictions the time for adjudicators to make their decision only commences upon the expiry of the period within which the respondent is to deliver its response to the adjudication application. The adjudicator is obliged to reject any adjudication response that is not lodged within the prescribed period, which in Australia is usually ten working days. Furthermore, the legislation usually provides that the respondent in the adjudication cannot rely upon any ground in its response that was not set out in its response to the payment claim notice.

Whilst the NSW Act does require the claimant to give the respondent a second opportunity to provide a response to the payment claim[34] (called a payment schedule), the consequences of the respondent failing to do so are severe to the extent of being described by one commentator as representing a deficiency in procedural justice.[35] Under section 22(2) adjudicators are confined in making their determination to considering the following matters only:

a   the provisions of the Act;
b   the provisions of the construction contract from which the application arose;
c   the payment claim to which the application relates together with the claimant's submissions;
d   the payment schedule if any; and
e   the results of any inspection carried out by the adjudicator.

In *Brookhollow Pty Limited v R&R Consultants Pty Limited*[36] the plaintiff sought to injunct the enforcement of the adjudicator's decision in circumstances where a payment schedule had not been delivered. Under subsection 20(2A) *Brookhollow* was precluded from lodging an adjudication response. As a result, in dealing with the matter, the adjudicator could only have regard to the items listed above and could only, in effect, see one side of the case. The principal argument put forward by *Brookhollow* was that the payment claim made by *R&R* was invalid, because it was not served within the period required by the legislation and, therefore, the adjudicator lacked jurisdiction. The court dismissed the application to injunct enforcement of the judgment obtained on foot of the adjudicator's decision, stating that had a payment claim notice not been served at all, the machinery would not have been set in motion and *R&R* could not rely upon the adjudicator's decision. However, a plea to the effect that it had not been served in time was 'like a defence in bar'.[37] It was a defence that the respondent could plead through a payment schedule, but it was not a defence that it could raise seeking, effectively, to set aside the decision.

As to the adjudicator's obligation under section 22(2) to have regard to the payment claim to which the application relates, the court found that the adjudicator did not have to go any further than to test whether: 'It purported reasonably on its face to comply with section 13(2)' (the requirements as to the substance of the claim) and not to enquire as to whether 'it also complied with the requirements of section 13(4) and (5)', which dealt with the time for service and the obligation to not serve more than one payment claim in respect of each payment date under the construction contract.[38]

Forum shopping

What if the appointed adjudicator is not liked by the referring party? Can the referring party abort the adjudication by failing to refer the dispute to that adjudicator within seven days and subsequently commence the procedure again? This precise issue was dealt with by the Court of Appeal in *Lanes*

Group Plc v Galliford Try Infrastructure Limited.[39] Following the appointment of the adjudicator, the referring party, in the hope of having another adjudicator appointed more to its liking, allowed the time for service of documents to lapse, served a new Notice of Adjudication and applied to the ICE for another adjudicator to be appointed. The respondent objected to this conduct and initially refused to take part in the proceedings. It ultimately sought to set aside the adjudicator's decision, inter alia on the grounds that a party can only refer a dispute once to adjudication. Whilst condemning the practice of forum shopping, the court saw obvious logistical problems with a finding of the nature sought by the respondent and upheld the validity of the second adjudicator's appointment. This is further vindication of the decision of the Irish legislature not to follow the UK precedent and instead to have one panel administered by one central body.

Possible Amendment to the Legislation

The respondent only receives the documents at the same time as the adjudicator. To reduce the risk of ambush, it might be more appropriate if the respondent were served with the documentation to be relied upon at the time of service of the Notice of Intention. This would require the referring party to have decided at that stage the documents upon which it would be relying and, in effect, therefore, to have prepared its Referral Notice at the same time as the Notice of Intention. In practice, these documents would almost invariably be prepared at the same time and therefore this would not give rise to any hardship. On the other hand, the extra seven days or so could be very important to the respondent in an ambush scenario. It might be appropriate to consider at some stage a change in the law so that all the information and documentation now required by the referral would have to be provided with the Notice of Intention.

Subsection 6(6): Twenty-eight day time limit for decision

> (6)   *The adjudicator shall reach a decision within 28 days beginning with the day on which the referral is made or such longer period as is agreed by the parties after the payment dispute has been referred.*

It is to be noted that the 28-day period runs from the date of the referral under subsection (5) rather than from the date of the appointment of the adjudicator.

There is no limit as to the complexity or extent of a dispute that may be referred to adjudication. The period of 28 days obviously causes, or can cause, considerable difficulty for the adjudicator in:

a   affording both parties a reasonable opportunity to present their cases;
b   considering all the material provided at the outset, and subsequently, by the parties;
c   preparing a reasoned decision.

This is an extremely tight time restraint. The time can be extended with the consent of the claimant by a further 14 days, and indefinitely with the consent of both parties. However, the claimant is not obliged under any circumstances to give its consent to any extension of time.

The legislation does not provide for specific procedures and therefore parties in the UK have a habit of bombarding adjudicators with further material, whether it is requested or not, right up to the last moment of the last day. It is submitted that under the Irish Act, an adjudicator will be perfectly entitled to put a time limit on the making of submissions and the provision of information in accordance with paragraph 26 of the Code of Practice. The difficulty for the adjudicator usually arises by reason of the parties insisting upon responding to any new piece of information, thus creating a perpetual circle. Provided adjudicators make it clear that they are not going to consider, or possibly even read, any information provided after a certain time, neither party can justifiably complain provided a reasonable opportunity has been afforded to each party up to that point to make and respond to submissions. However, adjudicators must reserve the right to seek further comment or information from the parties at any time and must be alert to affording the parties an opportunity to deal with new issues arising. Otherwise, the procedure may not accord with natural justice.

Whilst in theory a dispute could arise that is incapable of being dealt with fairly within a period of 28 days, it is to be noted that in the UK no decision of an adjudicator has been set aside by the courts for this reason.

Adjudicators must be extremely wary when having regard to late information. Natural justice requires that they do not take account of significant information, which might materially affect their decision, without giving the other party an opportunity to respond to it (*Construct Interiors NZ Limited v Peter William Jones and KMB Interior Contracts Limited*)[40]. Adjudicators, however, cannot be faulted for not taking account of information where they have made it clear that they do not wish for further information to be furnished. Provided they do not take account of unsolicited information provided out of time, there is no need to afford the other party an opportunity to respond to it. This, of course, is subject to both parties having been given an opportunity prior to that point of responding to issues raised by its opponent.

This approach may not be valid in the UK. Paragraph 17 of the UK Scheme provides: 'The adjudicator shall consider any relevant information submitted to him by any of the parties to the dispute and shall make available to them any information to be taken into account in reaching his decision'.

On the face of it, this would appear to oblige adjudicators to take account of information, which was unsolicited, even if they had forbidden the parties from providing further submissions beyond a certain time. This could cause difficulty for adjudicators when submissions are received after they have drafted their determination and there is little or no time left for making amendments. However, the courts in the UK have tended to apply this provision loosely. The Court of Appeal in *Carillion Construction v Devonport Royal Dockyard*[41] approved the words of Jackson J in the Court of First Instance at page 84:

If an adjudicator declines to consider evidence which, on his analysis of the facts or the law, is irrelevant, this is neither (a) a breach of the rules of natural justice nor (b) a failure to consider relevant material which undermines his decision on Wednesbury grounds or for breach of paragraph 17 of the Scheme. If the adjudicator's analysis of the facts or the law was erroneous, it may follow that he ought to have considered the evidence in question. The possibility of such error is inherent in the adjudication system. It is not a ground for refusing to enforce the adjudicator's decision.

There is a marked contrast between the legislation applicable in Australia and that applicable in the United Kingdom and Ireland in this respect. The latter jurisdictions simply provide for the claimant lodging a referral and the adjudicator then making a decision within 28 days. Very little guidance is given to adjudicators as to how they should go about this task. The Australian legislation requires a formal response to be made to the claimant's claim within a prescribed time limit. If it is not received within that time limit, the adjudicator must proceed without it. The legislation generally provides for the adjudicator having a specific period of time from the date the response was received (or was to have been received) and generally appears to anticipate that adjudicators will base their decisions normally upon those two documents. For instance, the Queensland legislation at section 25(4), having provided for the making of an adjudication application by the claimant and an adjudication response by the respondent, states:

For a proceeding conducted to decide an adjudication application, an adjudicator–

a   may ask for further written submissions from either party and must give the other party an opportunity to comment on the submissions; and
b   may set deadlines for further submissions and comments by the parties; and
c   may call a conference of the parties; and
d   may carry out an inspection of any matter to which the claim relates.

The UK and Irish models lend themselves to an endless stream of submissions with an obligation, certainly in the UK, for the adjudicator to take account of all submissions received. The Australian model is far more structured. The claimant makes its application, the respondent responds to that application and the adjudicators decision is made subject to the adjudicator's entitlement to seek further submissions. The parties themselves do not have an entitlement as of right to make further submissions, and it would appear that the adjudicator does not have to take account of responses a claimant may wish to make, say to the respondent's response. Because the respondent is not entitled to raise any new matters not already covered by its response ('payment schedule') to the initial claim, the respondent in its response is only elaborating on a position, which had already been set out and of which the claimant is aware. That the claimant would not have an automatic right of comment in relation to the response is demonstrated by the amendments made to the Queensland legislation. The 2014 Act entitles the claimant to respond to any new issues arising out of the response document in relation to complex claims. The Act provides for the respondent being entitled to go beyond what was contained in the payment schedule only in respect of complex claims (a claim over $750,000, a claim in relation to a latent condition or a time-related claim). Therefore, a right of reply is provided for in the legislation only to this limited extent. The adjudicator's time for making a decision runs in these circumstances from the date of the reply.

It is submitted that the more structured approach of the Australian legislation is less likely to give rise to issues of natural justice. Both parties know the case to a reasonable degree before the adjudication commences from the payment claim notice and the payment schedule served in response. Both parties have an opportunity to elaborate on their case in the adjudication, and if the adjudicator can then proceed to make a decision without calling for further submissions they can hardly be faulted for doing so, albeit the distinction between what is elaboration and what is new can sometimes be blurred. In Ireland and in the UK, where the legislation does not provide for the procedure to be followed or for the dispute arising necessarily from a payment claim notice, adjudicators need to be constantly vigilant to ensure that each of the parties has been given every opportunity to respond to every new point of significance.

The Queensland Act at section 26(2), sets out the 'only' matters to be considered by adjudicators in making their decision. These include all submissions 'properly made' by the claimant and the respondent. Submissions properly made are those within the adjudication application and the adjudication responses provided for by the legislation, and those additional submissions specifically requested by the adjudicator. All other submissions are regarded as unsolicited and would not usually be regarded as properly made. The Supreme Court of Queensland in *J. Hutchinson v Cada Formwork*[42] decided that a properly made submission is one made in a manner provided for in the Act. It also decided that where adjudicators considered something that is not a properly made submission, they must ensure that the other party has an opportunity to respond to it. Generally speaking, adjudicators tend to check unsolicited submissions and decide whether or not to have any regard to them. It is probably warranted that the adjudicator would at least read the material to the extent of ascertaining the point being made. It may be an important point, such as an objection by the party to a submission made by the other party on the basis that it contained privileged documentation.[43] The bias of the legislation, however, is against submissions being made outside of those expressly provided for in the legislation.

The Malaysian Act is somewhat different to all others in two respects. First, it provides at section 11(1) for the claimant having a right of reply to the respondent's response to the adjudication notice. Second, adjudicators have the luxury of a period of 45 working days from the date of the final submission to make their decisions.

The Northern Territory of Australia Act also contains a unique provision in this regard. Under section 34(3)(a) the registrar of the central administration body overseeing adjudication in that Territory may consent to an extension of time for adjudicators to make their decision, without the consent of the parties.

The Queensland Amendment Act of 2014 is innovative in many respects and not least in this respect. It provides at section 25B that, if in the opinion of the adjudicator, the claimant and the respondent attempted but failed to agree upon a time extension, adjudicators may, if the dispute is a complex one (within the meaning of the legislation), of their own accord extend the time for the making of their decisions by up to five business days.

That the validity of the adjudicator's decision depends upon it being issued within the timeframe provided by the legislation, is reinforced by the recent decision of the TCC in *KNN Coburn LLP v GD City Holdings Limited*.[44] However, the case also emphasises the fact that a failure to object to the timeframe proposed by the adjudicator may imply agreement to an extension of time for the issuing of the determination. In that case, KNN sent a copy of the Referral Notice to the adjudicator on 31 January 2013, but indicated that the hard copy with supporting documentation would follow by courier. The hard copy was received on 1 February 2013 and the adjudicator counted the 28 days from that date, indicating in his timetable that his decision would be made on 1 March 2013.

The respondent raised no objection to the timetable, but when the decision was issued, it challenged the decision on the grounds that the Referral Notice was in fact issued on 31 January. The court concurred with the view that the Referral was received on 31 January on the grounds that the adjudicator was able to identify the issues in dispute without reference to the hard copy documents, which were to follow. However, it held that the respondent acquiesced in the extension of time to 1 March by reason of its failure to raise any objection to the adjudicator's timetable. Accordingly, the determination was enforceable.

As to the calculation of the commencement and end of the 28-day period, and other periods of time set out in the Act, section 18(h) of the Interpretation Act 2005 is relevant:

> Where a period of time is expressed to begin on or be reckoned from a particular day, that day shall be deemed to be included in the period and, where a period of time is expressed to end on or be reckoned to a particular day, that day shall be deemed to be included in the period.

It is assumed that the Irish courts will follow the line of reasoning in *Ritchie Brothers (PWC) Limited v David Philip (Commercial) Limited*[45] in holding that the making of a decision within the 28-day period (or such extended period as may be agreed) is a condition precedent to the validity of the decision. In *Simons Construction Limited v Aardvark Developments Limited*[46] the court had found that the date for completion of the decision (albeit the adjudication was under a form of contract and not under the UK Scheme, but applying the logic of the UK Scheme), was not fundamental to the validity of the decision. That judgment was overruled by the Court of Session in the *Ritchie* case and a number of subsequent TCC judgments have applied the reasoning of the *Ritchie* case. Given that the Irish legislation is almost verbatim with the UK Act of 1996 on this issue, it is most unlikely that the Irish courts would take a dissimilar view to that of the courts in the UK.

The requirement that adjudicators make their decisions strictly within the 28 days was robustly explained by Judge Peter Coulson in *AC Yule & Son Limited v Speedwell Roofing & Cladding Limited*:[47]

> Finally, I am of the view that, in a speedy process like adjudication, the need for certainty is paramount. I consider that the UK Act of 1996 reflects that in numerous ways. That certainty would be lost, it seems to me, if the 28 days was no longer regarded as a clear and mandatory requirement but merely a guideline. Equally, certainty would also be lost if an adjudicator was given as long as he wanted to provide an enforceable decision, provided only that the parties could not show clear prejudice as a result of any delays beyond the 28 days, or the agreed extended period . . . the benefits of speed and certainty underpin the statutory requirement that the decision of the adjudicator shall be provided within 28 days (or any extended period that is agreed), and not there-after. This makes it important that both the UK Act of 1996, and the Scheme are construed purposively to ensure that these objectives are maintained.

The approach to such time limits is universal in the adjudication world. As indicated, legislation in the non-European countries usually provides for a response being made to the referral document, and the time for adjudicators to make their decision would normally run from the date of the response, or, if none is delivered, from the latest date the response should have been delivered. Thus, for example, section 23(1) of the Australian Capital Territory Act provides: 'the adjudicator for an adjudication application must not decide the application until after the end of the period within which the respondent may give an adjudication response'. It goes on at subsection 23(3) to state:

> [t]he adjudicator must decide an adjudication application as soon as possible but not later than . . . 10 business days after the earlier of (i) the date on which the adjudicator receives the adjudication response; and (ii) the date on which the adjudication response is required to be given to the adjudicator under section 22.

Time limit for notification of the decision

The obligation on the adjudicator is to reach a decision within 28 days (or such extended period as may be applicable), but not necessarily to publish it within that period. Paragraph 19(3) of the UK Scheme provides that: 'As soon as possible after he has reached a decision, the adjudicator shall deliver a copy of that decision to each of the parties to the contract'.

In *Cubbitt Building & Interiors Limited v Fleetglade Limited*[48] the adjudicator made his decision within the 28 days, but did not deliver it to the parties until just after noon on the following day. With some reluctance, the court came to the conclusion that the delay was permissible and that the decision was binding. The court indicated, however, that normally, having regard to modern means of communication, a decision should be delivered within hours of the decision being made and put in writing.

There is no corresponding provision to paragraph 19(3) of the UK Scheme in either the Irish Act or the Code of Practice. The legislation, therefore, is silent as to the obligation of adjudicators to communicate their decision to the parties. This begs the question as to whether adjudicators are entitled to insist upon payment of their fees prior to the release of their decision. The relevant legislation elsewhere in the world provides expressly for this situation. For instance, the Singapore Act at section 17(1B) requires the adjudicator, where a response has been made, to determine the adjudication application within 14 days of the response. It does not say how or when the decision/determination is to be communicated to the parties. Section 31(2) says that adjudicators will not be entitled to any fee if they fail to make a determination on the application within the prescribed time, unless this is because the application is withdrawn or the dispute is settled. Subsection 31(3), however, says that subsection (2) will not apply in circumstances where adjudicators require payment of their fees and expenses before issuing their decision.

The Tasmanian Act at section 24(1) requires the adjudicator, if a response has been delivered to the application to determine the application, as soon as practicable and in any event within ten business days of receipt of the response. Subsection 37(4) states that adjudicators are not entitled to be paid fees or expenses in connection with the adjudication if they fail to determine the application within the time prescribed. However, that subsection does not apply if adjudicators refuse to notify the parties of their determination until their fees or expenses have been paid.

The Tasmanian and Singapore legislation are very similar in this respect, and both are typical of the legislation in the non-European countries. The adjudicator is obliged to determine the dispute within the time prescribed, but the legislation is silent as to when and how the decision is to be communicated. However, it is expressly provided that adjudicators will be entitled to be paid their fees and expenses if the publication of their decision is delayed pending payment of those fees and expenses. The clear intent is that adjudicators are not obliged to release their decision until their fees are paid, and any delay thereby caused will not affect the validity of the determination.

In Australia it is open to debate whether a failure to make or notify a decision within the time required would invalidate the adjudicator's decision. A leading commentator suggests that the same principle should apply in Australia as applies in the UK. However, he opines that in the light of the decisions of the New South Wales Court of Appeal and Supreme Court respectively in *Transgrid v Siemens Limited*[49] and *Brodyn Pty Limited (t/a Time Cost and Quality) v Davenport*[50] it is unlikely that a court would strike down a late adjudication determination.[51]

As to the entitlement of an adjudicator in Ireland to maintain a legitimate lien in relation to their fees, see page 129 below.

Subsection 6(7): Power to extend the period to 42 days

(7) *The adjudicator may extend the period of 28 days by up to 14 days, with the consent of the party by whom the payment dispute was referred.*

The research carried out by Glasgow Caledonian University[52] indicates that 49 per cent of disputes decided by adjudicators in the year ending April 2014 were decided within 28 days and a further 31 per cent were decided within 42 days. The remaining 20 per cent were decided within such extended period as the parties may have agreed. A determination by an adjudicator outside of the time period provided by statute or agreed by the parties would be invalid.

Recent amending legislation in Queensland recognises that in adjudication, a uniform set of rules for both relatively simple and complex disputes is not necessarily appropriate. Some time limits are extended to cater for claims in excess of $750,000, or where the claim relates to a latent condition or is time-related. For example, the time is extended for responding to the initial payment claim from 10 to 15 business days, and the time within the adjudication process for responding to the claim is extended from 5 business days to 10 business days.

Subsection 6(8): The adjudicator must act impartially

(8) *The adjudicator shall act impartially in the conduct of the adjudication and shall comply with the code of practice published by the Minister under section 9, whether or not the adjudicator is a person who is a member of the panel selected by the Minister under section 8.*

This requirement is also dealt with in the Code of Practice. The subject is discussed more fully below with reference to paragraph 23 of the Code of Practice.

There is no process in the legislation for challenging the appointment of an adjudicator on the basis of possible bias or otherwise. Where the perception of impartiality ends and the failure to apply the tenets of natural justice begins, can be blurred. To hear one party in the absence of the other would offend against natural justice. It is clear that the courts in England take a similar view to the obligations of an adjudicator in this respect as they do in relation to those of an arbitrator, notwithstanding that the adjudicator has the statutory entitlement to initiate fact finding.

Subsection 6(9): Power of the adjudicator to take the initiative

(9) *The adjudicator may take the initiative in ascertaining the facts and the law in relation to the payment dispute and may deal at the same time with several payment disputes arising under the same construction contract or related construction contracts.*

Adjudicator's initiative in ascertaining the facts and the law

The fact that the adjudicator may take the initiative in ascertaining the facts and the law does not entitle adjudicators to make decisions on the basis of their own knowledge of the facts or the law, unless the parties have had an opportunity of considering these views. It is well established that the rules of natural justice apply to adjudication,[53] albeit some allowance has to be made for the timeframe in which the decision has to be made. Two recent decisions of the Technology and Construction Court give rise to an interesting comparison on the extent to which adjudicators may be permitted to base their decisions on issues not specifically raised by the parties. In *ABB Limited v BAM Nuttall Limited*[54] Akenhead J found that the adjudicator had overstepped the mark, and refused to enforce his decision. In *CG Group Limited v Breyer Group Plc*[55] the same judge found that the adjudicator had not done so.

The starting point in any discussion of this nature is that the policy of the courts is to uphold and support adjudication and adjudicators' decisions. In the *ABB* case Judge Akenhead commenced his judgment by quoting from Chadwick LJ in *Carillion Construction Limited v Devonport Royal Dockyard Limited* as follows:

85   The objective which underlies the Act and the statutory scheme requires the Courts to respect and enforce the adjudicator's decision unless it is plain that the question which he has decided was not the question referred to him or the manner in which he has gone about his task is obviously unfair. It should only be in rare circumstances that the Court will interfere with the decision of the adjudicator. The Courts should give no encouragement to the approach adopted by DML in the present case; which . . . may, indeed, aptly be described as 'simply scrabbling around to find some argument, however tenuous, to resist payment'.

86   It is only too easy in a complex case for a party who is dissatisfied with the decision of an adjudicator to comb through the adjudicator's reasons and identify points upon which to present a challenge under the labels 'excess of jurisdiction' or 'breach of natural justice'. It must be kept in mind that the majority of adjudicators are not chosen for their expertise as lawyers. Their skills are as likely (if not more likely) to lie in other disciplines. The task of the adjudicator is not to act as arbitrator or judge. The time constraints within which he is expected to operate are proof of that. The task of the adjudicator is to find an interim solution which meets the needs of the case. Parliament may be taken to have recognised that, in the absence of an interim solution, the contractor (or sub-contractor) or his sub-contractors will be driven into insolvency through a wrongful withholding of payments properly due. The statutory scheme provides a means of meeting the legitimate cash-flow requirements of contractors and their sub-contractors. The need to have the 'right' answer has been subordinated to the need to have an answer quickly. The scheme was not enacted in order to provide definitive answers to complex questions.

Indeed, it may be open to doubt whether Parliament contemplated that disputes involving difficult questions of law would be referred to adjudication under the statutory scheme; or whether such disputes are suitable for adjudication under the scheme. We have every sympathy for an adjudicator faced with the need to reach a decision in a case like the present.

Akenhead J also quoted from the TCC judgment in *Cantillon Limited v Urvasco Limited*,[56] where the approach of the court to this view of natural justice was summarised as follows:

> 57 From this and other cases, I conclude as follows in relation to breaches of natural justice in adjudication cases:
>
> a It must first be established that the Adjudicator failed to apply the rules of natural justice.
>
> b Any breach of the rules must be more than peripheral; they must be material breaches.
>
> c Breaches of the rules will be material in cases where the adjudicator has failed to bring to the attention of the parties a point or issue which they ought to be given the opportunity to comment upon if it is one which is either decisive or of considerable potential importance to the outcome of the resolution of the dispute and is not peripheral or irrelevant.
>
> d Whether the issue is decisive or of considerable potential importance or is peripheral or irrelevant obviously involves a question of degree which must be assessed by any judge in a case such as this.
>
> e It is only if the adjudicator goes off on a frolic of his own, that is wishing to decide a case upon a factual or legal basis which has not been argued or put forward by either side, without giving the parties an opportunity to comment or, where relevant put in further evidence, that the type of breach of the rules of natural justice with which the case of Balfour Beatty Construction Co Ltd v The Camden Borough of Lambeth was concerned comes into play. It follows that, if either party has argued a particular point and the other party does not come back on the point, there is no breach of the rules of natural justice in relation thereto.

Adjudicators may also be faulted if they apply a method of assessment that had not been canvassed by the parties. In *Herbosh-Kiere Marine Contractors Limited v Dover Harbour Board*[57] Akenhead J refused to enforce an adjudicator's decision where the adjudicator had availed of a different method of assessment when there was no dispute between the parties as to the correct method of assessment to be used. As it was clear from the notice of dispute that the method of assessment was not part of the dispute, the adjudicator was found to be going off 'on a frolic of his own' and to have both exceeded his jurisdiction and breached the rules of natural justice.

The judgment of the Outer House of the Court of Session in *Miller Construction (UK) Limited v Building Design Partnership*[58] would suggest that whilst adjudicators may not be entitled to go on frolics of their own, neither are they necessarily limited by the submissions of the parties. The defendant sought to resist the enforcement of the adjudicator's decision on the basis that whilst the plaintiff had only made its case in negligence, the adjudicator had found that the defendant had a contractual responsibility and found against it on this basis. Lord Malcolm held that the adjudicator's decision should be upheld and in doing so observed:

> It is well established in the authorities that an adjudicator is given considerable leeway, and that the court should be very slow to refuse enforcement on the grounds of breach of natural justice or that the adjudicator exceeded his jurisdiction. For example, reference can be made to the decision in the case of Diamond (cited earlier), and to the judgment of Chadwick LJ in Carillion Construction Limited v Devonport Royal Dockyard Limited [2006] BLR 15 at paragraphs 52/3 and 84/7. Amongst other things it was confirmed that an adjudicator is not required to adopt one or other of the parties' submissions. He can take an intermediate position without giving notice of his intention to do so.

In so far as subsection 6(9) allows the adjudicator to take the initiative in ascertaining the facts and the law in relation to the dispute, the wording is taken almost verbatim from paragraph 13 of the UK Scheme. Whereas the Irish legislation through the Code of Practice and the UK legislation through the Scheme set out in positive terms the powers of the adjudicator, the legislation of the non-European countries does not expressly entitle the adjudicator to take the initiative, and this may be indicative of a different philosophy applicable in the two regions. The legislation of the European countries does not provide for specific procedures after the referral is made, this being left entirely to the adjudicator. The legislation of the non-European countries invariably provides for a written response being made by the respondent and sometimes for a reply to that response. The non-European countries, therefore, anticipate something more structured and more akin to legal pleadings, and encourage the role of the adjudicator being limited to the consideration of these documents and such further submissions, if any, that might arise. The legislation of the southern hemisphere tends to set out the 'only' matters to which the adjudicator is to have regard and to expressly state that the adjudicator will not have regard to issues which were not raised by the respondent in its response to the pre-adjudication payment notice/application.

The adjudicators appointed under Section 8 of the Irish Act are professional people experienced in dispute resolution within the construction industry. They will be entitled to draw upon their experience and expertise at least to some extent, without having to invite comment from the parties. In *J & E Davy t/a Davy v Financial Services Ombudsman*[59] the Ombudsman drew certain conclusions from his own experience as to the extent to which the committees of credit unions would be expert in the matter of financial investments. The Supreme

Court was asked to set aside the Ombudsman's decision on several grounds, including the fact that the Ombudsman drew upon his own experience in this regard. The Supreme Court, whilst acceding to the application on other grounds, made no criticism of the Ombudsman for calling upon his own experience in this manner. At paragraph 116, Finnegan J, delivering the unanimous decision of the court observed:

> I note that ss 57BI and 57BK, dealing with the appointment of the respondent and a Deputy Ombudsman respectively provide that the Council shall appoint in each case a suitably qualified person, presumably a person with knowledge and experience of the financial services sector. It must be presumed that in each case in carrying out their functions they will avail of such knowledge and experience and if they do so any decision they reach would not automatically be condemned as a breach of fair procedures. In the present case I am satisfied that there was no breach of fair procedures the respondent having reached his conclusion on the evidence before him.

The powers conferred on adjudicators are more fully dealt with in Chapter 12, but in a general sense it is worth noting that the powers conferred in the non-European countries are somewhat more restrictive. For instance, there is no mention in the Queensland Act, which is typical of the region, of adjudicator's being entitled to engage experts to assist them in the adjudication.[60]

Multiple disputes

As to the entitlement of the adjudicator to deal at the same time with other payment disputes, the equivalent provision in paragraph 8 of the UK Scheme under the UK Act of 1996 provides:

1   The adjudicator may, with the consent of all the parties to those disputes, adjudicate at the same time on more than one dispute under the same contract.
2   The adjudicator may, with the consent of all the parties to those disputes, adjudicate at the same time on related disputes under different contracts, whether or not one or more of those parties is a party to those disputes.

One does not find similar provisions in the legislation of the non-European countries and there is really no clue to be gleaned from paragraph 8 of the UK Scheme as to what the intention of the Irish legislation was in this respect. In effect, the UK Scheme says that an adjudicator may not adjudicate at the same time on more than one dispute under the same contract and may not adjudicate at the same time on related disputes under different contracts, unless all of the parties concerned so agree. Subsection 6(9) of the Irish Act seems to say exactly the opposite, i.e. that an adjudicator may adjudicate ('deal with') several disputes arising under the same contract or under related contracts, whether or not the parties are so agreeable. There are two distinct elements to this aspect of sub-section 6(9). Firstly there is the entitlement of an adjudicator to deal with a number of different disputes under the same contract or related contracts. Adjudicators therefore may be appointed, for example, to deal with sub-contract disputes in circumstances where they have already been appointed in respect of the main contract. Secondly, there is the entitlement to deal with all such disputes "*at the same time*". The inability of adjudicators in the UK to deal with separate disputes at the same time has caused difficulties for adjudicators in that jurisdiction. In *Deluxe Art & Theme Limited v Beck Interiors Limited*[61] an adjudicators decision was not enforceable in circumstances where he had been appointed as adjudicator in respect of a number of different disputes between the same parties through separate adjudications, because the final adjudication overlapped with one of the other adjudications and the consent of the parties to the adjudicator dealing with the disputes at the same time had not been obtained.

Section 108 of the UK Act entitles a party to refer 'a dispute' to adjudication. The jurisprudence in that jurisdiction has defined the entitlement as referring to a single dispute, albeit that term is widely defined so that the manner in which the dispute is described can be very important. If a party challenges in adjudication a certificate as to the value of the final account, this would be regarded as a single dispute albeit it may have multiple aspects. On the other hand, a dispute as to whether two unrelated events give rise to an entitlement to payment may not be regarded as 'a dispute'. Ramsey J in *Willmott Dixon Housing Limited v Newlon Housing Trust*[62] speculated in *obiter dictum* that:

> an argument based on the reference in Section 108(1) to 'a dispute' being 'one dispute' may not be correct and that the reference to 'a dispute' is more likely to be a generic reference to 'a dispute', without seeking to limit it to a singular dispute.

In reaching this conclusion the court had regard to the Scheme, which allows more than one dispute to be dealt with by the adjudicator where the parties so consent. As their consent cannot override the provisions of the Act itself, the judge concluded that a wider interpretation was appropriate. However, the law in the UK is currently to the effect that only a single dispute can be referred at any time to adjudication, albeit the single dispute may involve numerous issues or elements. Coulson J in the *Deluxe Art* case suggested a useful test for deciding whether a dispute is a single dispute or not:

> A useful if not invariable rule of thumb is that, if disputed claim No 1 cannot be decided without deciding all or parts of disputed claim No 2, that establishes such a clear link and points to there being only one dispute.

Subsection 6(1) of the Irish Act read in conjunction with subsection 6(9) would suggest that the Irish legislation did not intend to confine the entitlement to refer a 'payment dispute' to a 'single dispute'.

Subsection 6(13): Correction of clerical errors

(13) *The adjudicator may correct his or her decision so as to remove a clerical or typographical error arising by accident or omission but may not reconsider or re-open any aspect of the decision.*

A similar provision to this was introduced in the UK by the UK Act of 2009. It was held in *Bloor Construction (UK) Limited v Bowmer & Kirkland (London) Limited*[63] that, in fact, a construction contract should be construed as implying a term of this nature. If that decision was correct, it is arguable that there was no need for an amendment to the UK Act of 1996. It is obviously appropriate, however, that the legislature in Ireland should expressly cater for this requirement.

Subsection 6(13) does not impose any time limit under which the adjudicator may make the correction. In the *Bloor* case the court expressed the view that the implied right of adjudicators to correct their determination would have to be exercised in a reasonable timeframe. Nonetheless, the UK Act of 2009 did not impose any time limit on the adjudicator. This is consistent with legislation elsewhere. For instance, the Singapore Act, containing a similar entitlement to correct mistakes, is also not subject to any time limit.[64]

Questions may arise as to whether an erroneous application of the slip rule gives rise to an issue of jurisdiction, in which case the correction will be invalid, or an error in law or fact, in which case it would not. Some assistance may be found in the judgment of Ramsey J in the case of *O'Donnell Developments Limited v Build Ability Limited*:[65]

> Secondly, if the adjudicator is asked by one party to correct a slip and he accepts that an error has been made within the slip rule then, if the adjudicator makes an error of fact or law in so doing, I consider that such an error does not take the exercise of the slip rule outside his jurisdiction. Finally, if the adjudicator is asked by one party to correct a slip, which the other party agrees is a slip within the slip rule but in operating the slip rule he makes an error of fact or law, then I do not consider that the court can interfere in that decision.

These words are consistent with the general approach of the courts to adjudication in the UK showing a preference not to interfere with the adjudicator's decision by reason of errors of fact or law. Provided the error is made within jurisdiction, the courts will enforce the defective decision.

Legal representation during the adjudication process

Some of the non-European countries discourage the involvement of lawyers in the adjudication process. Under the Tasmanian Act, it is specifically provided that conferences with the adjudicator are to be: 'conducted informally and may not be attended by a legal representative of any party' (section 24(5)). Under the

New Zealand Act, parties are expressly entitled to be represented by lawyers in any adjudication. Under section 16(5) of the Singapore Act, where an adjudicator has called for a conference of the parties to an adjudication, a party shall not be represented by more than two representatives (whether legally qualified or not) unless the adjudicator permits otherwise.

Whereas adjudicators in Ireland may very well discourage the involvement of lawyers in the adjudication process, it is most unlikely that they would seek to exclude them. When conciliation was first introduced into standard construction contracts in Ireland, some conciliators attempted to exclude lawyers, but over the 15 years or so since, conciliators have bowed to the inevitable. Anecdotally, it is understood that parties involved in adjudication in England are less inclined to avail of lawyers than is the case in Northern Ireland.[66] The Irish, both north and south, do seem to rely heavily upon lawyers in any form of dispute resolution.

Confidentiality

The legislation in some countries specifically provides for the confidentiality of the process. The Malaysian Act, for example, under the heading of 'Confidentiality of Adjudication' provides:

> The adjudicator and any party to the dispute shall not disclose any statement, admission or document made or produced for the purposes of adjudication to another person except:-
>
> (a) With the consent of the other party;
> (b) To the extent that the information is already in the public domain;
> (c) To the extent that disclosure is necessary for the purposes of the enforcement of the adjudication decision or any proceedings in arbitration or the court; or
> (d) To the extent that disclosure is required for any purpose under this Act or otherwise required in any written law.

Most of the legislation relating to adjudication, however, is silent on this issue. The common law implies a term of confidentiality into arbitration agreements,[67] but that arose from the fact that parties voluntarily enter into arbitration. It is unlikely that a term of confidentiality would be inferred from a statutory process. Therefore, if confidentiality is required, this is a term to be written into the contract between the parties.

The Department of Public Expenditure and Reform considered the possibility of making public adjudicators' decisions, but ultimately the Code of Practice included at paragraph 37 a confidentiality provision whereby: 'any document or information supplied for and/or disclosed in the course of the adjudication will be kept confidential'. It is not clear whether these words would include the adjudicator's decision. Is that decision to be regarded as a "document": 'disclosed in the course of the adjudication'? Presumably it is.

The position in the UK as to whether adjudication is subject to an implied term of confidentiality is unclear. Most of the rules, such as those of the Technology & Construction Bar (TECBAR), provide for confidentiality but, absent such rules, the issue has not been tried before the courts. Pending resolution of the issue, parties are likely to treat the process as being a confidential one. Under the Code of Practice, documents or information supplied and/or disclosed in the course of the adjudication are to be kept confidential. As a result, contractors and subcontractors are not likely to pool information. Therefore, the state may be in the same privileged position in terms of having a pool of knowledge available to it on adjudicators, similar to that which it enjoys under the public works contracts in respect of arbitrators. This is not a healthy state of affairs.

In Queensland adjudicators' decisions are made public. Anyone can access any decision made by any adjudicator through the website of the Building and Construction Industry Payment Agency.

Other aspects of confidentiality are discussed in Chapter 12 with reference to paragraph 37 of the Code of Practice.

Section 9: Code of Practice for adjudication

The Minister may prepare and publish a code of practice governing the conduct of adjudication under section 6.

The Code of Practice is discussed below at Chapter 12.

Section 10: Delivery of notices, etc.

(1) *The parties to a construction contract may agree on the manner by which notices under this Act shall be delivered.*

(2) *If or to the extent that there is no such agreement, a notice may be delivered by post or by any other effective means.*

(3) *Where under this Act a notice is required to be delivered not later than a specified number of days after a particular date and the last of those days is a day which is a Saturday or Sunday or a public holiday (within the meaning of the Organisation of Working Time Act 1997), the notice shall be taken to be validly delivered if delivered on the next day which is not such a day.*

Section 10 makes no effort to avoid an ambush. This could have been achieved to some extent by applying a wider scope to subsection 10(3) so as to avoid public holidays being counted where they fall inside the notice period, and by excluding from the notice period a number of days over the Christmas holiday period.

Subsection 10(1) is clearly open to abuse. A party drafting a contract might provide that service must be personal on a named individual, who may be difficult to locate.

References

1 [2008] EWHC 1020 (TCC); [2008] BLR 354; [2008] All ER (D) 106 (May); 119 ConLR 137, [2008] All ER (D) 106 (May)
2 Treasure & Son Limited v Martin Dawes [2007] EWHC 2420 (TCC); [2008] BLR 24; [2007] 44 EG 181 (CS), [2007] All ER (D) 386 (Oct)
3 *Fiona Trust & Holding Corporation v Privaler & Others* [2007] 2 All ER (Comm) 1053; [2007] UKHL 40, [2007] 4 All ER 951, [2007] Bus LR 1719, [2008] 1 Lloyd's Rep 254, 114 ConLR 69, (2007) Times, 25 October, 151 Sol Jo LB 1364, [2008] 4 LRC 404, [2007] All ER (D) 233 (Oct)
4 Civ 2005–404–5526
5 Civ 2009–404–3784, 5 August 2009
6 *Carroll (aminor) v Budget Travel Limited and Counihan Travel International*, Unreported High Court 7th December 1995
7 [2005] EWCA Civ 291; [2005] 1 WLR 2339, 101 ConLR 26, [2005] BLR 227, [2005] 12 EG 219 (CS), (2005) Times, 22 March, [2005] ArbLR 4, [2005] All ER (D) 280 (Mar)
8 e.g. *Collins (Contractors) Limited v Baltic Quay Management (1994) Limited* [2004] EWCA Civ 1757; 99 ConLR 1, [2005] 05 LS Gaz R 26, [2005] BLR 63, (2005) Times, 3 January, [2004] ArbLR 15, [2004] All ER (D) 97 (Dec)
9 [2009] EWHC 2890 (TCC); [2010] BLR 59; [2009] All ER (D) 240 (Nov)
10 [2012] EWHC 1808 (TCC); [2012] BLR 417, [2012] All ER (D) 31 (Jul)
11 *Fastrack Contractors Limited v Morrison Construction Limited* [2000] BLR 168; (2000) 75 ConLR 33, 16 Const LJ 273, [2000] All ER (D) 11;
12 [2004] EWCA Civ 1757; 99 ConLR 1, [2005] 05 LS Gaz R 26, [2005] BLR 63, (2005) Times, 3 January, [2004] ArbLR 15, [2004] All ER (D) 97 (Dec)
13 [2009] EWHC 2425 (TCC); [2010] BLR 363
14 [2002] EWHC 400 (TCC); [2002] BLR 312; 82 ConLR 24, [2002] All ER (D) 325 (Mar)
15 [2008] BLR 250; [2008] EWHC 282 (TCC), 117 ConLR 1, [2008] All ER (D) 406 (Feb)
16 [2009] EWHC 64 (TCC); 123 ConLR 15, [2009] Bus LR D76, [2009] All ER (D) 240 (Jan)
17 One exception to this is the Northern Territory Act, which at section 33 expressly requires the adjudicator to dismiss the application without making a determination if they are satisfied it is not possible to fairly make a determination in such circumstances
18 Guidance Note: Jurisdiction of the UK Construction Adjudicator, second edition (12/2012), paragraph 2.45
19 [2005] EWCA Civ 1358; [2006] BLR 15; (2005) Times, 24 November, [2005] All ER (D) 202 (Nov). The Referral Notice ran to 67 pages and was accompanied by 7 lever arch files. The adjudicator was furnished ultimately with 29 folders and was allowed a period of ten weeks to consider and decide all the issues
20 *CIB Properties Limited v Birse Construction Limited* 2004] EWHC 2365 (TCC), [2005] 1 WLR 2252, [2005] BLR 173, [2004] All ER (D) 256 (Oct):

   In my view the test which the adjudicator set himself, namely that he could only reach a decision if (a) he had sufficiently appreciated the nature of any issue referred to him before giving a decision on that issue including the submissions of each party and (b) if he was satisfied that he could do broad justice between the parties, was impeccable.

   (Toulmin J)

21 Page 189 of the Wallace Report
22 *Mecright Ltd v T.A. Morris Developments Ltd* [2001] Adj. L.R. 06/22
23 [2004] EWHC 2439 (TCC); 97 ConLR 142; 148 Sol Jo LB 1314, [2004] All ER (D) 426 (Oct)
24 [2013] EWHC 798 (TCC); 147 ConLR 194, [2013] 2 EGLR 23, [2013] BLR 325, [2013] All ER (D) 42 (Apr)
25 *Connex South Eastern Limited v MJ Building Services Group Plc* 2005] EWCA Civ 193, [2005] 2 All ER 870, [2005] 1 WLR 3323, 100 ConLR 16, [2005] BLR 201, (2005) Times, 13 May, 149 Sol Jo LB 296, [2005] All ER (D) 14 (Mar)
26 [2008] EWHC 1020 (TCC); 119 ConLR 137, [2008] BLR 354, [2008] All ER (D) 106 (May)
27 [2000] BLR 272; (2000) 70 ConLR 1, [2000] Lexis Citation 3222, 16 Const LJ 366, [2000] All ER (D) 559
28 [1843–60] All ER Rep 378; (1843) 3 Hare 100, 67 ER 313, 1 LTOS 410
29 [1981] 147 CLR 589; 36 ALR 3, 55 ALJR 96
30 [2000] 5 LRC 223
31 [2008] EWHC 3434 (TCC)
32 [2007] BLR 30; [2006] EWHC 2857 (TCC), 109 ConLR 67, [2006] All ER (D) 232 (Nov)
33 [2013] EWHC 2879 (TCC); [2013] All ER (D) 33 (Oct)
34 S.17(2)
35 Comparative Review of Construction Industry Payment Legislation and Observations from the Australian Experience by Jeremy Coggins, The International Construction Law Review, volume 29, page 214
36 [2006] NSWC SC 1
37 Paragraph 46
38 Paragraph 76
39 [2011] EWCA Civ 1617; [2012] Bus LR 1184, 141 ConLR 46, [2012] BLR 121, [2012] 13 Estates Gazette 92,[2011] All ER (D) 179 (Dec)
40 HK AC Civ 2010–404–897
41 [2005] EWCA Civ 1358; 104 Con LR 1; [2006] BLR 15; (2005) Times, 24 November, [2005] All ER (D) 202 (Nov)
42 [2014] QSC 63
43 Adjudication Decision 5153 by Debora Wardle, 1 July 2015, *Grindley Construction Pty Limited and Parbrock Pty Limited*
44 [2013] EWHC 2879 (TCC); [2013] All ER (D) 33 (Oct)
45 2005] CSIH 32, (2005) Times, 24 May, 2005 SLT 341, 2005 SCLR 829, 2005 Scot (D) 42/3
46 [2003] EWHC 2474; [2004] BLR 117; 93 ConLR 114, [2003] All ER (D) 482 (Oct)
47 [2007] EWHC 1360 (TCC); [2007] BLR 499, [2007] All ER (D) 100 (Jul)
48 [2006] EWHC 3413 (TCC); [2006] 110 ConLR 36; [2007] All ER (D) 268 (Jan)
49 [2004] 61 NSWLR 521; 21 BCL 273; [2004] NSWCA 395
50 [2003] NSWSC 1019
51 Marcus Jacobs, Security of Payment in the Australian Building & Construction Industry, fifth edition, page 606
52 Glasgow Caledonian University, Adjudication Reporting Centre, Report No 13, October 2014.
53 *Balfour Beatty Construction Limited v Lambeth London Borough Council* [2002] BLR 288; [2002] EWHC 597 (TCC), 84 ConLR 1, [2002] All ER (D) 60 (Apr); *Discain*

Project Services Limited v Opecprime Developments Limited [2002] BLR 402; *Carillion Construction Limited v Devonport Royal Dockyard Limited* [2005] EWCA Civ 1358, 104 Con LR 1, [2006] BLR 15; (2005) Times, 24 November, [2005] All ER (D) 202 (Nov)

54 [2013] EWHC 1983 (TCC); 149 ConLR 172, [2013] BLR 529, [2013] All ER (D) 224 (Jul)

55 [2013] EWHC 2959 (TCC)

56 [2008] EWHC 282 (TCC); 117 ConLR1; 2008 BLR 250; [2008] All ER (D) 406 (Feb)

57 [2012] EWHC 84 (TCC); 140 ConLR 97, [2012] BLR 177, [2012] All ER (D) 187 (Jan)

58 [2014] CSOH 80

59 [2010] 3 IR 324; [2010] IESC 30

60 The powers of the adjudicator are set out at section 25(4) of the Queensland Act

61 [2016] EWHC 238 (TCC); 164 ConLR 218, [2016] BLR 274, [2016] All ER (D) 125 (Feb)

62 [2013] EWHC 798 (TCC); 147 ConLR 194, [2013] 2 EGLR 23, [2013] BLR 325, [2013] All ER (D) 42 (Apr)

63 [2000] BLR 314; [2000] Lexis Citation 1251

64 Building and Construction Industry Security of Payment Act 2004, Section 17(6)

65 [2009] EWHC 3388 (TCC); [2009] 128 Con LR 141

66 A talk given by Edward Quigg to the Adjudication Society in Belfast in 2014

67 *Ali Shipping Corp v Shipyard Trogir* [1999] 1 WLR 314; [1998] 2 All ER 136, [1998] 1 Lloyd's Rep 643

7 Selection of panel of adjudicators

Subsection 8(1): Selection by the Minister

(1) *The Minister shall from time to time select persons to be members of a panel (in this section referred to as the 'panel') to act as adjudicators in relation to payment disputes and shall select one of those persons to chair the panel.*

Currently, the majority of main contracts in Ireland are between state funded authorities and main contractors. The form of main contract in use is unique to Ireland and is blatantly biased in favour of the employer. One of the provisions in that contract requires that the Minister for Finance be provided with a copy of every arbitrator's award issued under these contracts. This enables the Minister to acquire a unique knowledge of the interpretations put on provisions in the contract by individual arbitrators, which in turn could be used to inform the Minister's selection of an arbitrator under those contracts. The fact, therefore, that the Minister is to appoint a panel of adjudicators and is to appoint the chair of that panel is somewhat disturbing, given that the state is likely to be directly or indirectly a party to a large percentage of the adjudications that will take place. However, the appointment of Dr Nael Bunni as the Chair is seen by the industry as a step in the right direction and has gone a long way to allay any concerns in this regard.

The minister named in the Act is the Minister for Public Expenditure and Reform. As that minister is also the paymaster, or the ultimate paymaster, in respect of state funded construction projects, it was in hindsight thought by the Government to be inappropriate that the Minister for Public Expenditure and Reform should be the party selecting persons to be members of the panel. Therefore, by Statutory Instrument 476 dated 14 October 2014 the Minister for Public Expenditure and Reform has been replaced by the Minister for Jobs, Enterprise and Innovation.

Arbitration in the UK is a confidential process by way of an implied term of the arbitration agreement.[1] It is almost certain that the Irish Courts would follow that precedent. However, it is not a breach for a party to advise the Minister of the outcome of an arbitration and to provide a copy of the arbitrator's determination to him, because the form of public works contract allows for this. It would, however, be a breach on the part of the main contractor to notify the Construction Industry Federation or others as to the outcome of the arbitration and, as stated, this potentially gives the employer an advantage over the contractor in the selection of an arbitrator. The Code of Practice renders the adjudication process also confidential, but nonetheless certain information is to be provided by the adjudicators to the chair. Potentially, therefore, a similar advantage might arise in favour of the state in respect of adjudication. It may be important, to safeguard confidence in the process, that systems be put in place to ensure that information provided under paragraph 39 of the Code of Practice is not made available to government agencies other than the chairperson and the construction contracts adjudication service.

Subsection 8(2) provides that a party selected to be a member of the panel will remain a member for five years and shall be eligible for reselection at the end of the five-year period. Subsection 8(3) allows the Minister for good and sufficient reason to remove a member from the panel, and subsection 8(4) allows a member to resign at any time by giving notice to the Minister. Obviously, more detailed rules will have to be established if the panel of adjudicators is to command the respect of the industry. It is essential that members of the panel are seen to be people of the utmost integrity and competence. It is likely that parties, in seeking to agree an adjudicator, will work from the panel in selecting an appropriate appointee. Checks and balances therefore have to be put in position to ensure that those admitted to the panel remain competent throughout the period they are on the panel. Panel members should be required to keep themselves entirely abreast with the development of adjudication and the law relating to it. At the very least, therefore, adjudicators should be required to provide CPD certification of a high standard not less than once a year. The Wallace Report,[2] in recommending the abandonment of the practice of appointments by nominating bodies in favour of a central register operating under a government appointed registrar, recommended a number of steps to be taken to ensure high standards, such as the following:

1   The panel of adjudicators should be divided into active and non-active adjudicators. In Queensland a considerable number of lawyers are qualified as adjudicators but never act as such. Non-active adjudicators would not be appointed by the registrar unless they applied to be put on the list of active adjudicators and were found to be suitable to make the transfer.
2   The appointment of adjudicators should be made by an independent public servant who will have regard to such matters as the size of the claim, the nature and complexity of the claim, whether there are issues of a specialist nature involved, etc.
3   An independent body should assess allegations of misconduct against an adjudicator.
4   Any active adjudicator must advise the registrar in writing within one business day of their engagement on behalf of any party to an adjudication.

The latter was a controversial proposition albeit a compromise on the view expressed elsewhere in the Report:

> The Review is aware that some adjudicators act as claims preparers. Indeed it has been submitted to the Review, that if adjudicators were not permitted to also act as claims preparers, many would be forced to leave the industry in search of other work. That is not in my view adequate justification for maintaining the status quo.

In Ireland the construction industry is a small one, and those engaged in construction alternative dispute resolution are small in number. Those who are engaged as conciliators and arbitrators tend also to represent parties in arbitrations and conciliations. It is likely that those who are engaged as adjudicators will also wish to advise parties and, if this were not the case, the chances are that the pool of adjudicators remaining would be so small in number as to be unhealthy. The fact, however, that adjudicators are likely to be involved in advising clients should be recognised by the chair at least to the extent of introducing stringent requirements for adjudicators in making declarations of interest.

Subsection 8(5) sets out the persons who will be eligible to be members of the panel. In all cases the Minister is to have regard to the candidate's experience and expertise in dispute resolution procedures under construction contracts. Panel members will be one of the following:

a   a quantity surveyor or building surveyor registered under the Building Control Act 2007;
b   a chartered member of the Institute of Engineers Ireland;
c   a barrister;
d   a solicitor;
e   a fellow of the Chartered Institute of Arbitrators;
f   a person with qualifications equivalent to any of the above in any other member state of the EU.

The question arises as to whether the parties may provide in their contract for the adjudicator to be appointed, in default of agreement, by some third party other than the chair of the Minister's panel, such as Engineers Ireland or the Royal Institute of Architects of Ireland. The Irish Act provides at section 6(4) that failing agreement between the parties upon the adjudicator to be appointed, the adjudicator shall be appointed by the chair of the panel. Would a term that failing agreement the appointment is to be made by some such third party be a limitation or exclusion of a provision of the Irish Act and therefore be void by reason of subsection 2(5) of the Irish Act? Prior to the publication of the Code of Practice in July 2016, it seemed certain that a provision in a contract providing that a specific person would act as an adjudicator in the event of a dispute, would be void. Paragraph 8 of the Code of Practice has cast some doubt on this. It is thought a provision whereby the adjudicator would be appointed by some institution other than the chair of the panel would be void, but this is not obvious. There is a clear implication to be drawn from subsection 6(3) that if an individual is to be agreed, this can only happen after the notice of adjudication and therefore any attempt to appoint an individual through the contract cannot succeed. However, the legislation does not, by necessary inference or otherwise, prohibit the appointment being made by an institution other than the chair of the panel. If the parties agree in the period contemplated by section 6(3) that they will abide by a nomination by a third party, the appointment might be seen as an appointment made under that sub-section and therefore as a valid appointment. On the other hand, if the parties through their contract decide that the adjudicator is to be nominated by a particular third party and that the parties will be bound to appoint the nominee, this may not be valid.

The legislation of most countries is more explicit either through the primary legislation or regulations as to what is required by way of qualifications. For instance, the Singapore Building and Construction Industry Security of Payment Regulations 2005 require that in addition to having an appropriate professional degree or diploma, the adjudicator has: 'working experience of at least ten years in, or relating to, the building construction industry in Singapore, and has successfully completed the pre-qualification assessment and training course conducted by the authorised nominating body'.[3]

The relevant professional bodies in Ireland have organised and held rigorous courses in adjudication for those with a high level of experience and expertise in construction industry dispute resolution. One might have expected that the successful completion of such a course would be a necessary pre-qualification to being placed on the Minister's panel or, if not a pre-qualification, that the fact that a person passed such a course would automatically admit them to the panel. Because there is no reference to any such qualification criterion, obviously such a qualification, through either this or any other course, cannot be a prerequisite to admission to the panel. Beyond that, the extent to which the Minister is permitted to take account of such qualifications in selecting the panel is unclear. Applicants are entitled to be admitted to the panel if they meet the criteria set out in the Irish Act. Whereas the fact that a person has passed a particular course may establish that person's knowledge of adjudication and expertise in dispute resolution procedures, a person applying for admission would clearly be entitled to establish that expertise through other means.

Neither the Irish Act nor the Code of Practice require that the adjudicator has any training in adjudication or qualifications in adjudication. Given that this requirement is not set out in the primary or secondary legislation, the question arises as to whether the Minister is entitled to insist upon such training or qualifications without amending the legislation.

The Collins Report suggested that in New South Wales there should:

> [b]e instituted a more intensive and detailed training course to be successfully completed before any person can qualify to act as an adjudicator . . .
>
> Adjudicators training and refresher courses should be devised and conducted by an independent neutral and competent body qualified to do so, such as the Institute of Arbitrators and Mediators Australia.

The Queensland Act at section 18 requires that a person eligible to be appointed as an adjudicator in relation to a construction contract must have: 'successfully completed a formal course of training of at least two days duration in adjudication of payment disputes'. The Wallace Report pointed out that the Queensland legislation (unlike the Irish Act) does not require an adjudicator to hold an academic qualification. The report suggested no change be made in that regard:

> I do not think it is appropriate to shut the gate on persons who may otherwise make very good adjudicators because they do not have a formal qualification from a tertiary institution. Whilst it is unlikely that a non-tertiary educated adjudicator may be called upon to decide a multimillion dollar adjudication application, I consider it is very important for the integrity of the Act and for the betterment of the industry to permit the registration of trade based adjudicators for example.
>
> It must be remembered that over the past two years, between 75% and 78% of adjudication applications involved payment claims of less than $66,000. Often, a law degree or a degree in architecture or quantity surveying for example is not required. What is required is an independent person who having obtained their adjudication qualification, drawing upon their own industry experience has the requisite skills to be able to decide the value of a payment claim.

Recent reforms introduced in Queensland move away from the claimant choosing a nominating institution to a centralised appointing body similar to the chair appointed under the Irish Act. The benefits are considered to be twofold. First, adjudication in Australia was falling into disrepute, because claimants were choosing nominating bodies perceived to be claimant friendly. The Society of Construction Law (Australia) Report of June 2014 is particularly revealing and outspoken on this issue.[4] Second, there is a cost saving in that a state controlled body is likely to charge a smaller fee than an institution for the appointment.

The Irish Act and the Code of Practice are silent on the issue of how the panel is to be administered other than saying that it will be administered by a chair appointed by the Minister. One of the issues to be considered is whether the chair should be concerned to ensure that the fees charged by adjudicators are uniform and, if not uniform, are reasonable. Another is whether adjudicators are to be appointed on a taxi rank system, or based on a particular adjudicator being appointed by reason of their skill set. The Wallace Report recommends against the taxi rank system, but suggests that the appointing body, to counter any perception of favouring one adjudicator over another, should be required to publish details of the number of applications referred to each adjudicator during a relevant period.[5]

In August 2015 the Department of Jobs, Enterprise & Innovation invited applications for inclusion on the Ministerial appointed panel of adjudicators. Some interesting issues have arisen from the invitation. First, there is an emphasis on geographical location. Paragraph 3 of the information booklet indicates that applicants may during the selection process be requested to select locations in respect of which they would be willing to conduct cases. It goes on to state that the chair will endeavour to allocate the cases to the adjudicators appointed to the panel based on their expressed preference for geographical location.

Apart from being bound by the Code of Practice, a previous draft of the Code of Practice provided that adjudicators appointed to the panel would also be required to abide by a Code of Conduct: 'relating to the behaviour of adjudicators . . . and may be required to sign a document agreeing to this requirement'. It is understood that there will be a Code of Conduct for members of the panel albeit this is not now a requirement of the Code of Practice.

Paragraph 12.6 of the information booklet for applicants proved to be controversial:

> A member of the Panel, for the duration of their membership of the Panel, shall not advise or provide representation to any party in a case where the Chairperson of the Construction Contracts Adjudication Panel has appointed a member of the Panel of Adjudicators to the case.

This provision was presumably intended to discourage active lawyers or claims consultants and advisors from applying to be members of the panel. There are very few arbitrators or conciliators who make their living solely from that activity. Most are engaged in advising clients in relation to disputes when not acting as an arbitrator or conciliator. One cannot say how many experienced professionals who might otherwise have been admitted to the panel did not apply for membership by reason of this provision.

As indicated, Section 8(1) requires the Minister to: 'select persons to be members of a panel and to "select" one of those persons to chair the panel'. The Minister appears to have operated in the reverse order to that required by the Act. The Minister appointed Dr Bunni as the chair (or the 'chair designate') to the panel, prior to selecting the panel. Potentially, this could affect the validity of appointments made by the chair.

A panel of thirty adjudicators has now been selected and published. A number of relatively young applicants (under 40) were not even granted an interview notwithstanding that they had passed the rigorous conversion course run by the barrister Peter Aeberli and organised jointly by the relevant professional bodies. This presumably was because they were considered not to have the requisite experience required under the legislation. The panel comprises of five engineers, six architects, six lawyers, and thirteen quantity surveyors[6]. Noticeably the panel does not include one lady. There are in fact very few active conciliators/arbitrators who are female and it is thought that very few women applied to become members of the panel.

Notwithstanding the contents of paragraph 12.6 of the rules applicable to any application, a small number of active lawyers, intending to continue to advise clients in relation to adjudication, were appointed. It is understood that the chair of the panel will not in fact seek to enforce paragraph 12.6. It is a fact nonetheless that this paragraph did prevent other lawyers applying for membership of the panel.

References

1 *Ali Shipping Corp v Shipyard Trogir* [1998] 2 All ER 136; [1999] 1 WLR 314, [1998] 1 Lloyd's Rep 643
2 Final Report of the Review of the Discussion Paper – *Payment Dispute Resolution in the Queensland Building and Construction Industry* by Andrew Wallace, B. L., 24 May 2013
3 Paragraph 11(1B)
4 Report on security of payment and adjudication in the Australian construction industry by the Australian Legislation Reform Sub-Committee of SCL
5 The Wallace Report, page 164
6 Many of the panel have dual qualifications. For the purpose of counting the numbers, the author has had regard to the member's primary profession

8 Extent to which adjudicator's decision is binding

Subsection 6(10): Adjudicator's decision

(10) *The decision of the adjudicator shall be binding until the payment dispute is finally settled by the parties or a different decision is reached on the reference of the payment dispute to arbitration or in proceedings initiated in a court in relation to the adjudicator's decision.*

The purpose of adjudication is to ease the cash flow of the executing party. Adjudication provides a swift method of obtaining a neutral party decision as to whether monies claimed are payable or not. The purpose of the exercise would be defeated if the other party could postpone the application of the decision of the adjudicator by simply referring the dispute to arbitration or litigation as the case may be. Under this subsection the parties are obliged to implement the decision of the adjudicator pending the outcome of any other proceedings. In effect, this will usually mean a payment being made to the claimant. In practice the parties tend to implement the decision of the adjudicator and not seek to have the dispute dealt with in arbitration or in court.

It is perhaps alarming to recall that the first draft of the Scheme in the UK anticipated that the adjudicator's decision would be final and binding for all time – there would be no right of appeal or fresh hearing of the issue through litigation or arbitration. A large body of opinion supported that proposition. It was largely as a result of the intervention of Lord Ackner in the House of Lords on 22 April 1996[1] that a consultative process on this provision was undertaken. This ultimately led to the UK Scheme being delayed for two years and to the finality of the adjudicator's decision being subject to the outcome of arbitration and litigation.

All other countries introducing adjudication have followed suit. As the whole purpose of adjudication is to facilitate cash flow, the adjudicator's decision in all jurisdictions is binding pending the outcome of arbitration or litigation.

Subsection 6(12): Adjudicator's decision binding for all purposes

> (12) *The decision of the adjudicator, if binding, shall, unless otherwise agreed by the parties, be treated as binding on them for all purposes and may accordingly be relied on by any of them, by way of defence, set off or otherwise, in any legal proceedings.*

This subsection does not entitle the parties to opt out of the Irish Act or of the provisions in the Irish Act whereby the adjudicator's decision is binding. Section 2(5) prevents the parties from limiting or excluding the application of the Irish Act. Therefore, an adjudicator's decision will always be enforceable provided it is valid. What subsection 12 allows is for the parties to limit by agreement the extent to which the adjudicator's decision can be relied upon in any proceedings other than those through which enforcement of the adjudicator's decision is sought. It is difficult to know what exactly is sought to be achieved through this sub-section. It may encourage parties through their contracts to find ways of limiting the applicability of adjudicator's decisions while at the same time remaining within the Act.

Adjudicator's decision binding whether right or wrong

It is well settled since *Bouygues (UK) Limited v Dahl-Jensen (UK) Limited*,[2] a Court of Appeal decision, that the courts will enforce an adjudicator's decision notwithstanding that the decision is clearly in error, whether that be an error of law or an error of fact. The facts of that case were both stark and startling. Because of an error in calculation, the adjudicator had allowed a sum of about £200,000 in favour of *Dahl-Jensen* instead of allowing *Bouygues* a sum of about £140,000. It was clear that the adjudicator had made an error of calculation in respect of retention and *Dahl-Jensen* did not seek to argue otherwise. Nonetheless, the court, having regard to the purpose of adjudication and the risk of rough justice entailed in achieving that purpose, was of the view that the adjudicator's decision must be upheld.

This principle has been applied regularly by the Technology and Construction Court and at this stage is a cornerstone of adjudication in the UK.

It is to be noted that the adjudicator in *Bouygues* was offered an opportunity to correct his mistake. However, he refused to recognise the error. It would appear from the judgment in *Bloor Construction (UK) Limited v Bowmer & Kirkland (London) Limited*[3] that an adjudicator had an entitlement to correct a clerical error provided this was done quickly. That entitlement arose out of an implied term of the construction contract. The *Bloor* decision was given some support by Dyson J in *Edmund Nuttall v Sevenoaks District Council*.[4] Regardless of whether the decision is right or wrong, the issue has been overtaken by an amendment to the UK Scheme whereby an adjudicator now has a statutory entitlement to correct errors of this nature. If, however, the adjudicator refuses to correct the error, no matter how obvious, the law in the UK requires the courts to uphold the decision. The Irish Act and the legislation of the non-European countries also permit the adjudicator to correct mistakes. This is expressed in many different ways, but the essence of the power is to correct accidental slips or omissions and mistakes of identification or form. The adjudicator is not entitled to correct an error of reasoning or judgment.

The facts in *Geoffrey Osborne Limited v Atkins Rail Limited*[5] were very similar to those in *Bouygues*. In it the adjudicator simply failed to make a deduction of £912,000 for monies that were already payable under a certificate to the contractor. The result was that he found that the contractor was entitled to a further payment in excess of £500,000 instead of finding that the contractor owed the employer a sum in excess of £400,000. Again there was no dispute that the adjudicator had made an error. The employer *Atkins Rail Limited* was able to overcome the error in the particular case by asking the court to make a finding by way of declaration, which would be final and binding on the parties in the sense of being a final determination by legal proceedings under Section 108(3) of the UK Act of 1996. The court was not in a position to rehear the entire of the issue that was before the adjudicator, but considered itself in a position to make a decision on a specific aspect, the effect of which was to negate the error made by the adjudicator:

> 80   I consider that, in the light of my conclusions in this judgment, ARL is entitled to a declaration to the effect that the adjudicator was wrong to order payment of sums to GOL in respect of his assessment of the value of its claims in payment application No 36 without taking into account and, if appropriate, deducting from the amounts assessed, any sums that ARL had paid or allowed to GOL by the date of the Notice of Adjudication.

It is to be noted that this was a rare instance where the contract between the parties did not provide for arbitration. Had it done so, the court could not have negated the effect of the error, because Section 108(3) of the UK Act of 1996 provides:

> The contract shall provide in writing that the decision of the adjudicator is binding until the dispute is finally determined by legal proceedings, by arbitration (if the contract provides for arbitration or the parties otherwise agree to arbitration) or by agreement.

Adjudicator's decision binding on certifier

An adjudicator's decision is not just binding in the sense that the parties are entitled to insist upon its application in terms of immediate payment. It is also binding on the certifier in respect of all future certificates unless and until the adjudicator's decision is overturned. This principle was clearly established in the case of *William Verry Limited v The Mayor & Burgesses of the London Borough of Camden*.[6] If it were otherwise, each successive certificate could be used to defeat an adjudicator's decision on a previous certificate, and the fundamental purpose of providing cash flow would be defeated. However, the court held that there was one exception to this. If new facts came to light of which the adjudicator had not been aware, those new facts could be taken into account by the certifier in any subsequent certificate. If, however, such facts were known at the time of the adjudication and were simply not argued before the adjudicator, those facts cannot be availed of to diminish the impact of the adjudicator's decision.

In the *William Verry* case Ramsey J pointed out that prior to the decision in *Ferson Contractors v Levoloux AT Limited*[7] there had been court decisions, which appeared to contemplate that a contractual right might allow a party to avoid in whole or in part the obligation to comply with the adjudicator's decision. In *Bovis Lend Lease Limited v Triangle Developments Limited*,[8] for example, the court had found that where other contractual terms clearly had the effect of superseding or providing for an entitlement to avoid or deduct from a payment directed to be paid by an adjudicator's decision, those terms would prevail.

Ramsey J considered that judgment was not in keeping with the intention of Parliament as reflected in *Ferson v Levoloux* or indeed as stated by Dyson J in *Macob Civil Engineering Limited v Morrison Construction Limited*:[9]

> The intention of Parliament in enacting the Act was plain. It was to introduce a speedy mechanism for settling disputes in construction contracts on a provisional interim basis and requiring the decision of adjudicators to be enforced pending the final determination of disputes by arbitration, litigation or agreement.

An adjudicator's decision merely providing that the executing party is entitled to payment on foot of a contract administrator's certificate, as opposed to a valuation carried out by the adjudicator, would not be binding in respect of future certificates nor future adjudications. If the proper construction of sub-section 4(3) of the Act is to the effect that the executing party is entitled to payment of the amount claimed in the event of the other party failing to deliver a response to the payment claim notice, neither would that decision be binding in respect of future certificates or future adjudications (*Rupert Morgan Building Services (LLC) Limited v Jervis*[10] and *Harding t/a MJ Harding Contractors v Paice & Another*[11]. In these circumstances the adjudicator is not making a decision as to the value of the work done but merely a decision as to the legal entitlement of the Claimant by way of interim measure. It is assumed that the Irish Courts will take the same approach to this issue as has been taken by the Courts of England and Wales albeit the courts in that jurisdiction have relied to some extent upon paragraph 9 of Part 1 of the Scheme, dealing with a situation where the dispute referred to the adjudicator is the same or substantially the same as one already decided in adjudication, and there is no equivalent provision in the Act or in the Code of Practice.

Adjudicator's decision binding on later adjudications

In so far as an adjudicator decides a point of principle, that decision is not only binding upon the parties and a certifier in respect of future certificates but is also binding upon the adjudicator. Once adjudicators have made a determination on a point of principle, they cannot change their mind in a later adjudication between the same parties. This was precisely the point at issue in *Vertase FLI Limited v Squibb Group Limited*.[12] This was a dispute between a main contractor and a sub-contractor. In the first adjudication, the adjudicator decided that the main contractor was not entitled to recover liquidated and ascertained damages against the sub-contractor unless it was obliged to pay such damages under the main contract. In the second adjudication, he was persuaded that he was wrong in reaching this conclusion and decided that the main contractor was entitled to liquidated and ascertained damages, notwithstanding that no such damages were visited upon it by the employer. The Technology and Construction Court found that the adjudicator was bound by his previous decision. The court took the view that since the parties were bound by the decision of the adjudicator in the first adjudication, that decision was final and binding between the parties unless and until it was determined otherwise by litigation or arbitration and: 'it matters not whether it is right or wrong'.[13]

A close analysis of the adjudicator's decision may sometimes be required to ascertain whether or not their decision incorporates a finding on a particular issue. In *KNN Coburn LLP v GD City Holdings Limited*[14] the Technology and Construction Court was asked not to enforce an adjudicator's decision where he ruled upon the liquidated damages to which *KNN* was entitled in circumstances where it was alleged that this had already been determined by a previous adjudicator. The previous adjudicator had decided: 'On the basis of my conclusions about GD City's entitlements to extension of time, KNN have a putative claim for liquidated damages in the sum of £20,000.00'.

However, *KNN* had not served a withholding notice and the earlier adjudication had found that the entitlement to deduct liquidated damages was not automatic, but relied upon notice being served under the contract between the parties. The adjudicator in so ruling had not made any decision as to *KNN*'s entitlement to liquidated damages. The adjudicator had merely found that no entitlement in respect of liquidated damages could be established in that adjudication, because no claim had been made in that regard by way of withholding notice.

The position in the UK on first sight appears to be clear as to the extent to which a later adjudicator is bound by a decision of an earlier adjudicator. Paragraph 9(2) of the UK Scheme copper fastens subsection 108(3) of the UK Act of 1996: 'An adjudicator must resign where the dispute is the same or substantially the same as one which has previously been referred to adjudication, and a decision has been taken in that adjudication'.

This concept, however, can be problematic in cases involving complex issues such as extension of time. In *Quietfield Limited v Vascroft Construction Limited*[15] the Court of Appeal was faced with circumstances where the contractor, *Vascroft,* had sought an extension of time in an earlier adjudication for the exact same period as was sought in a later adjudication. In the first adjudication the adjudicator had found that Vascroft had not proved its case. In the later adjudication Vascroft relied on some of the grounds on which it had previously relied, but on this occasion it also relied upon other grounds and furthermore produced a very detailed analysis of the impact of the delays on the critical path. In the Court of Appeal the court posed the issue as follows:

In the present proceedings, the judge said uncontroversially in para 29 of his judgment:

'It is clear from the particulars of claim, the defence and the skeleton arguments in this action that there is effectively only one issue to be decided. That issue may be formulated as follows: Was the adjudicator correct in treating his own decision in the first adjudication as conclusive in relation to extension of time? If the answer to this question is "Yes", then the adjudicator's decision dated 7 December 2005 must be enforced. If the answer to this question is "No", then it follows that the adjudicator has expressly refused to consider both the written submissions and the evidence which constitute Vascroft's only substantive defence in the adjudication. In that event there has been a breach of the rules of natural justice and the adjudicator's decision cannot be enforced'.[16]

The Technology and Construction Court (Jackson J) had gone on to conclude:

Whether, at the end of the day, the submission in Appendix C will prevail, I do not know. That would be a matter for the adjudicator or, possibly, the arbitrator to decide. I am, however, quite satisfied that Vascroft's alleged entitlement to an extension of time as set out in Appendix C is substantially different from the claims for extension of time which were advanced, considered and rejected in the first adjudication.

In the appeal the position of Quietfield was recorded at paragraph 28 as follows:

Mr. Holt pointed out that the Referral Document spoke of a 'comprehensive' extension of time claim. He submitted that Vascroft were building a picture and asking the first adjudicator to look at everything, whether it had been previously notified or not. They claimed a full extension of time to the 23rd September 2005. The scope of the first adjudication should be construed as embracing all claims for extension that might have been made. It is said that Quietfield so understood the scope of the first adjudication. They regarded the first adjudication as in substance determining, for the purposes of adjudication at the least, their entitlement to liquidated damages. There was nothing in Appendix C which could not have been put forward in the first adjudication.

The Court of Appeal dismissed the appeal. It agreed that the issues in the second adjudication were substantially different to the issues in the first adjudication.

Section 108(3) of the UK Act of 1996 provides: 'that the decision of the adjudicator is binding until the dispute is finally determined by legal proceedings, by arbitration . . . or by agreement'. The first adjudication had decided that Vascroft was not entitled to an extension of time on the grounds put forward. The first question is whether the grounds put forward are even relevant. If the issue is whether Vascroft was entitled to an extension of time and the adjudicator decided that it was not, is that not an end of the matter? The second question, if the adjudicator's decision is only binding to the extent of the application made in the first adjudication, is whether the adjudicator's decision was binding at least in respect of those grounds. Therefore, if the adjudicator decided that the specific grounds put forward had not been established as causing delay, was Vascroft not bound by that decision to the extent of not being able to put forward those grounds in a second adjudication? Apparently not. Appendix C in the second adjudication was a comprehensive document dealing in detail with the impact of the delays upon the critical path. May LJ concluded:

> It could well be that grounds for extension of time, which were not established individually in the first adjudication, could nevertheless legitimately feature in App C, in conjunction with other grounds not advanced in the first adjudication, as being on a critical path affected by those other causes of delay. In principle, such a composite claim might legitimately be seen as outside the dispute which the first adjudicator determined.

Dyson LJ was perhaps somewhat more circumspect in this regard. At paragraph 47 he stated:

> Whether dispute A is substantially the same as dispute B is a question of fact and degree. If the contractor identifies the same Relevant Event in successive applications for extensions of time, but gives different particulars of its expected effects, the differences may or may not be sufficient to lead to the conclusion that the two disputes are not substantially the same. All the more so if the particulars of expected effects are the same, but the evidence by which the contractor seeks to prove them is different.
>
> Where the only difference between disputes arising from the rejection of two successive applications for an extension of time is that the later application makes good shortcomings of the earlier application, an adjudicator will usually have little difficulty in deciding that the two disputes are substantially the same.
>
> In the present case, I am in no doubt that the judge reached the right conclusion. The first disputed claim which was the subject of the first adjudication was substantially different from the second disputed claim. The written notices which formed the basis of the second claim identified Relevant Events which were substantially more extensive than those which formed the basis

of the first claim. The particulars of expected effects were very different too. There will be some borderline cases where it is a matter of judgment whether the two claims are substantially the same and where there may be room for more than one view. In my view, this is not a borderline case.

The Court of Appeal accepted in this case that there was no reason why Vascroft could not have put forward its Appendix C submission in the first adjudication to ground its application for an extension of time for the same period as covered by the second adjudication. Arguably, therefore, the dispute was the same, and all that changed were the grounds and evidence to support the claim. The judgment of the Court of Appeal in *Quietfield Limited v Vascroft Construction Limited* has been applied in a number of other cases.[17]

In the UK and elsewhere, a claimant is entitled to payment of the full amount claimed in its application if the respondent fails to serve an appropriate counternotice under the legislation. The claimant in these circumstances is entitled to payment in full as an interim measure. However, the respondent is entitled to question the value of any interim payment in subsequent adjudications. In other words, the default entitlement is not binding for the purpose of future valuations, notwithstanding that it is the subject matter of an adjudicator's decision.[18] This is because the valuation of the amount claimed is not part of the adjudicator's reasoning. The adjudicator merely allows the amount claimed because that is what the law requires. Therefore certificates issued after the adjudicator's decision may value the work at less than the amount claimed and allowed by the adjudicator. This will presumably be the position in Ireland also if it is decided that the amount claimed in a payment claim notice has to be paid by the other party if that other party fails to respond in accordance with sub-section 4(3).

Set-off and counterclaim

This arises in two respects. A case for set-off or counterclaim may be made in the adjudication itself or it may be made by way of defence to an application to enforce the adjudicator's decision. As is evident from the *William Verry* case, the latter will rarely be allowed. To the extent that it may be allowed, this would usually arise where liquidated damages flow inexorably from the logic of the adjudicator's decision, albeit that decision does not provide for liquidated damages. The tendency of the courts in the UK, however, has been against allowing any deduction for liquidated damages (*Edmund Nutall Limited v Sevenoaks District Council*[19] and *RJ Knapman Limited v Richards*[20]).

In *ROK Building Limited v Celtic Composting Systems Limited*,[21] it was stated at paragraph 17:

> It is now well established law and practice in the context of construction adjudications that valid adjudicators' decisions are to be enforced in effect without set-off or cross claims. So far as set-offs or cross claims are concerned, the logic is that these are to be raised in the adjudication and the adjudicator

either allows them or disallows them; it is not then appropriate that the losing party raises on enforcement proceedings either the same set-offs or cross claims (which have already been adjudicated upon) or new ones which could have been but were not raised.

If the responding party in an adjudication believes that it has an entitlement to set off monies by way of counterclaim or by way of liquidated damages, it must raise these issues in the adjudication (provided it is not prevented from doing so by reason of a failure to comply with section 4(4)). If it fails to do so, it is most unlikely that any such right of set-off or counterclaim will be taken into account in an application to the court to enforce the adjudicator's decision. The responding party who fails to raise such issues will be entitled to refer those issues to adjudication and, if successful, to recover whatever is due on foot of the second adjudicator's decision. In the meantime, however, it has to pay on foot of the first adjudicator's decision.

Set-off of liquidated damages, however, may be permissible even outside of the adjudication where the only logical conclusion from the adjudicator's decision is that such liquidated damages are due, or in circumstances where the contract provides for the deduction of liquidated damages in circumstances that have clearly been met. This principle was set out by Jackson J in
Balfour Beatty Construction v Serco Limited[22] as follows:

> [39] I derive two principles of law from the authorities, which are relevant for present purposes:
>
> (1) Where it follows logically from an adjudicator's decision that the employer is entitled to recover a specific sum by way of liquidated and ascertained damages, then the employer may set off that sum against monies payable to the contractor pursuant to the adjudicator's decision, provided that the employer has given proper notice (in so far as required).
>
> (2) Where the entitlement to liquidated and ascertained damages has not been determined either expressly or impliedly by the adjudicator's decision, then the question whether the employer is entitled to set off liquidated and ascertained damages against sums awarded by the adjudicator will depend upon the terms of the contract and the circumstances of the case.

The decision of Coulson J in *Workspace Management Limited v YJL London Limited*[23] is an interesting one in that it contained a finding that the adjudicator had jurisdiction to make a decision that the adjudicator himself disowned. In most cases on jurisdiction the challenging party seeks a ruling that the adjudicator exceeded their jurisdiction. It is rarely that circumstances would give rise to the court considering whether the adjudicator had failed to exercise or realise their full jurisdiction. The adjudicator had in effect determined that there was no money due by the contractor to the developer, but had stopped short of saying that there was money due by the developer to the contractor by reason of a payment made by the developer on foot of a previous decision. The adjudicator confirmed in subsequent correspondence that he did not feel that he had jurisdiction to go that far. At paragraph 22 of the judgment the court stated:

> Miss Cheng submitted that the Adjudicator had the jurisdiction to reach a nil valuation of Certificate 27, but not the jurisdiction to go on and consider whether any sum was due to the Defendant from the Claimant. As she put it, 'Once he got to nil, he could stop'. I consider that that argument takes an unrealistic view of the valuation process that the Adjudicator went through. He carried out his detailed valuation. It was only when he totalled up the figures and compared the result with what had been already paid that he would have become aware that the sum due was less than the amount that had already been repaid by the Defendant to the Claimant. He could not do part of that valuation and then stop. It was a composite exercise and had to be completed in full.

On this basis, the court went on to conclude that the adjudicator did in fact have jurisdiction to find that a sum was due by the developer to the contractor. The court relied on a number of other points to confirm this view including:

> In Federal Commerce & Navigation Co Ltd v Molena Alpha Inc [1978] QB 927, [1978] 3 All ER 1066, [1978] 3 WLR 309, Lord Denning MR said:
>
> 'It is not every cross-claim which can be deducted. It is only cross-claims that arise out of the same transaction or are closely connected with it. And it is only cross-claims which go directly to impeach the Plaintiff's demands, that is, so closely connected with his demands that it would be manifestly unjust to allow him to enforce payment without taking into account the cross-claim'.
>
> [39] In my judgment, for the reasons that I have set out, it would be manifestly unjust to allow the Claimant to enforce payment without taking into account the cross-claim based on Adjudication Decision 3. The principles of equitable set-off are, in my view, triggered in the present case.

Most of the standard contracts in use in Ireland require that certain prerequisites be met before liquidated damages are deductible. For instance, the RIAI contract form, at clause 29, requires that an architect's certificate be issued confirming the amount of the liquidated damages payable. Where such prerequisites are not met, obviously a claim for liquidated damages cannot be set off unless the issue has been brought squarely within the adjudication and the adjudicator has allowed a set-off or such a set-off follows inevitably from the adjudicator's decision. However, under the Public Works Contracts operable in Ireland, it is likely that the employer will be entitled to deduct liquidated damages (and indeed other sums) directly from any finding of an adjudicator, as such a deduction is allowed by the terms of contract without certification from a third party.

References

1 Hansard 22 April 1996, Columns 989 – 990 (see Coulson on Construction Adjudication, second edition, paragraph 1.31)
2 [2000] BLR 49; (1999) 70 ConLR 41, [1999] Lexis Citation 3672, [1999] All ER (D) 1281
3 [2000] BLR 314; [2000] Lexis Citation 1251
4 [2002] HT 00 119
5 [2009] EWHC 2425 (TCC); [2010] BLR 363
6 [2006] EWHC 761 (TCC)
7 [2003] EWCA Civ 11; [2003] 1 All ER (Comm) 385; [2003] BLR 118; 86 ConLR 98, [2003] 05 EGCS 145, 147 Sol Jo LB 115, [2003] All ER (D) 172 (Jan)
8 [2002] EWHC 3123 (TCC), at paragraph 37; 86 ConLR 26, [2003] BLR 31, [2002] All ER (D) 155 (Nov)
9 [1994] 64 Con LR 1; [1999] BLR 93; [1999] Cill 1470; [1999] 3 EGLR 7, [1999] 37 EG 173, (1999) Times, 11 March, [1999] All ER (D) 143
10 [2003] EWCA Civ 1563; [2004] 1 WLR 1867; [2004] 1 All ER 529, 91 ConLR 81, [2003] NLJR 1761, [2004] BLR 18, (2003) Times, 26 November, [2003] All ER (D) 153 (Nov)
11 [2015] EWCA Civ 1231
12 [2012] EWHC 3194 (TCC); [2016] 2 All ER 819, 163 ConLR 299, [2016] BLR 85, 166 NLJ 7680, [2015] All ER (D) 11 (Dec)
13 Edwards-Stuart J, paragraph 44
14 [2013] EWHC 2879 (TCC); [2013] All ER (D) 33 (Oct)
15 [2006] EWCA Civ 1737; 114 ConLR 81, [2007] BLR 67, [2007] Bus LR D1, [2006] All ER (D) 331 (Dec)
16 May LJ, paragraph 17
17 *Vertase FLI Limited v Squibb Group Limited* [2012] EWHC 3194 (TCC); [2013] BLR 352, [2012] All ER (D) 187 (Nov); *Carillion Construction Limited v Smith* [2011] EWHC 2910 (TCC), 141 ConLR 117, [2012] Bus LR D61, [2011] All ER (D) 121 (Dec); *HG Construction Limited v Ashwell Homes (East Anglia) Limited* [2007] EWHC 144 (TCC); 112 ConLR 128, [2007] BLR 175, [2007] All ER (D) 210 (Feb)
18 *Rupert Morgan Building Services (LLC) Limited v Jervis* [2003] EWCA Civ 1563; [2004] 1 WLR 1867; [2004] 1 All ER 529; 91 ConLR 81, [2003] NLJR 1761, [2004] BLR 18,(2003) Times, 26 November, [2003] All ER (D) 153 (Nov); *Galliford Try Building Limited v Estura Limited* [2015] EWHC 412 (TCC); 159 ConLR 10, [2015] BLR 321, [2015] All ER (D) 01 (Mar)
19 [2002] HT 00 119
20 [2006] EWHC 2518 (TCC); 108 ConLR 64, [2006] All ER (D) 349 (Oct)
21 [2009] EWHC 2664 (TCC); 130 ConLR 61; [2009] All ER (D) 65 (Nov)
22 [2004] EWHC 3336 (TCC) ; [2004] All ER (D) 348 (Dec)
23 [2009] EWHC 2017 (TCC); [2009] 3 EGLR 11, [2009] 42 EG 178, [2009] BLR 497, [2009] All ER (D) 119 (Aug)

9   Enforcement

Subsection 6(11): Leave of the High Court

(11) *The decision of the adjudicator, if binding, shall be enforceable either by action or, by leave of the High Court, in the same manner as a judgment or order of that Court with the same effect and, where leave is given, judgment may be entered in the terms of the decision.*

This provision is very similar to Section 23 of the Arbitration Act 2010:

An award . . . made by an arbitral tribunal under an arbitration agreement shall be enforceable in the State either by action or, by leave of the High Court, in the same manner as a judgment or order of that Court with the same effect and where leave is given, judgment may be entered in terms of the award.

In the case of arbitration, a judge has been designated under the Arbitration Act to deal with applications to the court, and this has ensured that such applications are given priority. Whilst the vast majority of adjudicators' decisions will be implemented by the parties without resort to court, it is considered essential for the proper functioning of the Irish Act that adjudicators' decisions are dealt with in the same way. Adjudication has been particularly effective in England, because of the robust support afforded to it by the Technology and Construction Court. The enforcement of adjudicators' decisions in Northern Ireland by the courts tends to be slow, and this does detract from the effectiveness of the legislation in that jurisdiction.

Statutory Instrument No 450 of 2016 incorporates the Rules of Court applicable to adjudication. The rules provide for an application being made by notice of motion grounded on affidavit. The rules require the responding party to deliver a replying affidavit within one week of the notice of motion being served and the applicant has a further week to respond to that replying affidavit. All of this is to occur before the motion date. In theory the court would then hear the motion. It is more likely that the motion will be assigned at that stage a date for hearing.

A similar provision to that contained in the Singapore Act might have been appropriate to deter unmeritorious resistance to enforcement proceedings:

> Where any party to an adjudication commences proceedings to set aside the adjudication determination or the judgment obtained pursuant to this section, he shall pay into the Court as security the unpaid portion of the adjudicated amount that he is required to pay, in such manner as the court directs or as provided in the Rules of Court pending the final determination of those proceedings.[1]

The New Zealand Act also provides an example of constructive provisions for recovery of both the amount claimed, where it becomes due without adjudication (section 23(2A)), and for recovery of the amount due on foot of an adjudicator's decision (section 59(2A)). These sections provide that the payee may recover not only the relevant amount in any court but also: 'the actual and reasonable costs of recovery awarded against the payer by that court'. The High Court in that jurisdiction has decided such costs are not subject to the usual rules of taxation. Provided the costs incurred are reasonable they are recoverable even if they include solicitor and own client costs as opposed to party and party costs.[2]

Obviously, the court will not enforce the adjudicator's decision where it is found that the decision is made without jurisdiction, or is significantly tainted by a failure to apply basic concepts of natural justice. In addition, the court has a discretion to put a stay of execution in appropriate circumstances. The TCC judgment in *Galliford Try Building Limited v Estura Limited*[3] provides an example of the court exercising similar discretion. In that case, *Estura* had, through an oversight, failed to serve a payment notice or a pay less notice with the effect that the contractor was entitled to payment of the full amount claimed in its application. When payment was not made the contractor sought to enforce the adjudicator's decision. At paragraph 53 Edwards-Stuart J found:

> Even though this is a situation which Estura has brought on itself by its failure to comply with the notice provisions in the contract . . . that does not mean that the court must refuse the grant of a stay of enforcement of any judgment irrespective of how unfair that might be to Estura.

In the unique circumstances of this particular case the Court found that not to grant a partial stay would give rise to a risk of 'irreparable prejudice'[4] to Estura. Given that the contractor had done nothing wrong, the court was anxious to avoid doing any injustice to the contractor by putting a stay on the entire of the adjudicator's determination. The court took account of a number of factors, including the fact that the contractor should not be put in any worse position than it would have been had the notices been served by Estura, and that Estura should not be put in any better position than would have been the case had the notices been served. This, of course, put the court in the difficult position of seeking to assess a sum of money without all the relevant evidence. The payment sought included a sum in excess of £2 million for loss and expense. The court decided to cut this sum by two-thirds stating: 'The starting point here is that experience shows that loss and expense claims are frequently significantly over valued, and that quite often the true value is about a third of the figure claimed'. Such an empirical assessment is obviously not ideal, but it was clearly a matter of the court doing its best in the circumstances. Even the enforcement procedure, therefore, may give rise to an element of rough justice.

The fact that under subsection 6(11) the Irish Act requires the leave of the court for the purpose of entering judgment on foot of an adjudicator's decision, closes the door on obtaining judgment through the High Court office as would be the norm in undefended debt collection proceedings. The fact that an application to the court is required will increase substantially the costs involved in adjudication. If the legislation is being amended at any future date, it may be worth considering a provision similar to that operating in New Zealand. Under the New Zealand Act a party seeking to obtain judgment must serve on the other party notice of the application. It is then for the other party to make an application to the court to prevent registration. If no such application is made, the claimant is entitled to mark judgment through the court office after the expiry of 15 working days.[5]

The starting point on the issue of enforceability is to recognise that as a matter of policy the courts in the UK are: 'hostile to technical arguments to postpone the enforcement of (the adjudicator's) decision' (*Carillion Construction Limited v Devonport at paras 85–87*).[6] In *Pioneer Cladding Limited v John Graham Construction Limited*[7] the court was faced with a situation where the successful sub-contractor in adjudication was clearly insolvent and would be unable to repay the monies payable under the adjudicator's decision if the court enforced that decision. The Technology and Construction Court was reminded of its earlier decision in the case of *Wimbledon Construction Company 2000 Limited v Derek Vago*[8] where the court summarised the relevant principles relating to the grant of a stay of execution in such circumstances as follows:

(a) Adjudication (whether pursuant to the UK Act of 1996 or the consequential amendments to the standard forms of building and engineering contracts) is designed to be a quick and inexpensive method of arriving at a temporary result in a construction dispute.

(b) In consequence, adjudicators' decisions are intended to be enforced summarily and the Claimant (being the successful party in the adjudication) should not generally be kept out of its money.

(c) In an application to stay the execution of summary judgment arising out of an Adjudicator's decision, the Court must exercise its discretion under Order 47 with considerations a) and b) firmly in mind (see AWG).[9]

(d) The probable inability of the Claimant to repay the judgment sum (awarded by the Adjudicator and enforced by way of summary judgment) at the end of the substantive trial, or arbitration hearing, may constitute special circumstances within the meaning of Order 47 rule 1(1)(a) rendering it appropriate to grant a stay (see Herschell).

(e) If the Claimant is in insolvent liquidation, or there is no dispute on the evidence that the Claimant is insolvent, then a stay of execution will usually be granted (see Bouygues[10] and Rainford House[11]).

(f) Even if the evidence of the Claimant's present financial position suggested that it is probable that it would be unable to repay the judgment sum when it fell due, that would not usually justify the grant of a stay if:

  (i) the Claimant's financial position is the same or similar to its financial position at the time that the relevant contract was made (see Herschel[12]); or

  (ii) the Claimant's financial position is due, either wholly, or in significant part, to the Defendant's failure to pay those sums which were awarded by the adjudicator (see Absolute Rentals).[13]

The issues arising out of the principle enunciated at (f)(i) are of particular interest. In the Pioneer case Coulson J came to the conclusion that the main contractor, John Graham Construction, sought to ascertain the claimant's financial position at the time of entering into the sub-contract and had been misled by the sub-contractor as to its assets and viability. On this basis, the court granted a stay of execution. Main contractors in Ireland do not usually conduct painstaking enquiries, as did Graham Construction in that case, into the financial viability of sub-contractors. Unless they change their habits in that respect, they may well find themselves in the position of having to pay the amount of adjudicators' decisions to insolvent sub-contractors without any hope of recovering these payments in the event of the adjudicators' decisions being found to be incorrect.

The fact that the defendant to the enforcement proceedings may itself have an adjudication pending, which may have the effect of cancelling or modifying the sum due to the plaintiff, is not of itself grounds for refusing enforcement (*Avoncroft Construction Limited v Sharba Homes (CN) Limited*[14] and *Interserve Industrial Services Limited v Cleveland Bridge UK Limited*[15]).

In *Galliford Try v Estura* the court took account of the severe financial hardship that would be imposed upon the paying party if it was obliged to discharge the full amount of the adjudicator's decision. In *Rmc Building & Civil Engineering Limited v UK Construction Limited*[16] Mr Justice Edwards-Stuart emphasised that the facts of the *Galliford* case were exceptional and that the refusal to enforce the adjudicator's decision in full is likely to arise very rarely stating at paragraph 56 of the judgment:-

The provisions introduced by the Act and the Scheme are all about maintaining cash flow. That purpose is not achieved by simply giving judgment for a sum and then staying its enforcement; interest is often no compensation for a lack of cash flow.

The cases referred to would suggest that the Courts in the UK may be more inclined to put a stay on part of the amount allowed by the adjudicator by reason of severe financial hardship if the adjudicator has allowed the full amount claimed because of the absence of a payment notice or pay less notice being served. This is because it would not be unusual that there would be a considerable element of exaggeration in the claim. However as emphasised in the *Rmc* case a party seeking to establish severe financial hardship must prove that circumstance and should not assume that the court will accept a mere assertion in that regard.

Jurisdiction

If a court refuses to enforce an adjudicator's decision it is usually because the decision is made without jurisdiction or unfair procedures have been applied. An adjudicator's decision made without jurisdiction is not binding provided, of course, that the error in jurisdiction was of significance. This statement may require further qualification in New South Wales in view of the decisions in *Brodyn Pty Limited t/a Time Cost and Quality of Davenport*[17] and *Chase Oyster Bar v Hamo Industries*,[18] but it is certainly the case in the UK, and there is little doubt that the Irish courts would take the same approach. Upon appointment, therefore, adjudicators must satisfy themselves that they have the jurisdiction to deal with the issues in dispute. This would mean checking carefully the issues in the referral with the issues identified in the Notice of Intention. Adjudicators derive their jurisdiction from the latter. The adjudicator's jurisdiction may also be affected by:

- The fact that there has been a previous adjudication between the parties and the dispute, or part of the dispute referred to the adjudicator has already been decided in an earlier adjudication.
- Whether the dispute is in fact a dispute relating to payment. That of course raises the thorny issue as to what exactly is meant by that expression.
- To what extent claims or counterclaims put forward by the respondent (whether mentioned or not in the response delivered by the respondent under subsection 4(4)) are within jurisdiction.
- Is the contract one to which the Irish Act applies?
- Has the adjudicator been appointed in accordance with the legislative requirements?
- Is there a dispute between the parties of the nature required by subsection 6(1)?

In some jurisdictions the legislation is restrictive as to claims or counterclaims that may be made by a respondent. Under the NSW Act (Section 20(2)(b)) a respondent may only make claims for set-off or counterclaims to the extent that it has included these in its response to the payment application. Legislation generally in the non-European countries provides expressly that the adjudicator will not take account of issues (including counterclaims) that have not been raised in response to the payment claim notice.

In the UK the respondent would usually be permitted to make its full case in the adjudication. Where, however, a withholding notice or a pay less notice is required, the paying party will be confined to such set-off or counterclaim as has been relied upon in those documents. The guidance notes published by the Adjudication Society jointly with the Chartered Institute of Arbitrators, third edition (July 2015) at paragraph 3.12 stated:

> A particularly common issue arises where the referring party claims a sum of money from the responding party which the responding party seeks to reduce by way of a counterclaim. The counterclaim will fall within the adjudicator's jurisdiction provided it satisfies the legal requirements of set off, which it generally will if it arises out of the same project. A failure to consider the counterclaim may render the decision unenforceable.

That statement is referenced in an earlier edition to the case of *Pilon Limited v Breyer Group Plc.*[19] This comment, however, must be read in context. In the *Pilon* case the claimant had primarily sought payment in relation to a specific aspect of its works. However, the notice of adjudication was interpreted by the court as including an application for all payments due and this entitled the respondent to raise a set-off in relation to defective work in another aspect. The general rule in the UK is that a counterclaim or unparticularised set-off cannot be raised if it was not included in a withholding notice. An issue could be raised, however, prior to the implementation of the 2009 Act by way of defence of set-off, as in the *Pilon* case, if the notice of adjudication was sufficiently wide as to embrace it. No defence of set-off would normally be entertained, however, where the payment sought was one certified under the contract, unless it had been expressly covered by a withholding notice.[20] This is because under the terms of the contract, the certified sum would be deemed to be due and payable.

It is important to note in the context of jurisdiction that the courts in the UK have refused to entertain arguments on jurisdiction where the issue could have been raised in the adjudication and was not, or where it was raised, but the party raising the issue nonetheless submitted to the adjudicator's jurisdiction.[21] If a party challenges the jurisdiction of the adjudicator and subsequently partakes in the adjudication whilst reserving its right to raise the issue again in the courts if necessary, it will not lose the entitlement to do so.[22] If, however, through its conduct it is seen to waive the point on jurisdiction, it will lose the entitlement to raise it later.

In practice, the claimant, faced with the dilemma of proceeding with the adjudication in the knowledge that the respondent is reserving the right to make its challenge on jurisdiction, might consider writing to the respondent inviting it to make an application to the court at that stage on the basis that the claimant will agree to extend the time for the making of the adjudicator's decision for a reasonable period after the court has made a determination on the issue of jurisdiction. In this manner, the claimant's dilemma may become that of the respondent. The respondent may then be put in the position of either making the challenge at that point or accepting the adjudicator's jurisdiction. The respondent could, of course, refuse to make the application to the court at that stage and seek to reserve its entitlement to do so at a later stage. However, the court's view on jurisdiction may be influenced by the Respondent's failure to accept this invitation. The issue could also be relevant to the adjudicator's determination as to the party liable for the payment of their fees and expenses.

Jacobs[23] points out that there is a marked difference between the approach by the courts in the UK and those in Australia to the question of jurisdiction. The Australian courts tend towards the view that adjudicators have the power to determine their own jurisdiction and provided that the determination is made in good faith, it will be upheld. This would be at odds with the approach of the courts in the UK. The justification for the Australian position is that this is what is required by the legislation. It is in the nature of a swift and inexpensive resolution of the dispute on a temporary basis that such errors as to jurisdiction will be forgiven. It remains to be seen which approach the Irish courts will take. Under the Arbitration Act of 2010 arbitrators are given an express power to decide upon their own jurisdiction. Prior to that legislation arbitrators had no such power. There is no provision in the Act allowing adjudicators decide upon their own jurisdiction and accordingly it is unlikely that they will be found to have such an entitlement.

Statute of limitations

A difficulty can arise where adjudication proceedings are dealt with many years after the original dispute arose. Suppose a contractor is dissatisfied with the final account. Under the UK Act it would be entitled to seek adjudication at any time. If the adjudicator's decision is given more than six years after the cause of action arose and the contractor succeeded in its claim, the employer may wish to have the issue referred to arbitration as a result of that decision, but may be statute barred by reason of the statutory time limit for issuing proceedings having expired.

The Technology and Construction Court found that time only begins to run under the statute of limitations in respect of an adjudicator's decision when the decision is implemented, i.e. payment is made (*Ringway Infrastructure Service v Vauxhall Motors No 2*).[24] This somewhat undesirable conclusion was thought to be logically necessary to protect the respondent in adjudication. If the time ran from any earlier date, the respondent would be unprotected in circumstances where the claimant commences adjudication near to the expiry of the time limit and a decision is made in favour of the claimant outside of the limitation period. If the time for instituting proceedings commenced at the time of the claimant's alleged breach, the respondent would be obliged, in advance of the adjudicator's decision, to issue proceedings for some form of declaration to protect the respondent against what the adjudicator might find. This could give rise to logistical and technical difficulties and would place an unfair burden on the respondent.

This decision was fully reviewed by the Court of Appeal in *Walker Construction v Quayside Homes Limited & Another*,[25] a judgment of 7 February 2014. The court compared it with the judgment of Lord MacFadyen in *City Inn Limited v Sheppard Construction Limited*,[26] with the decision of Akenhead in *Aspect Contracts (Asbestos) Limited v Higgins*[27] and that of Judge Davies in *Jim Ennis Construction Limited v Premier Asphalt Limited*[28] and came down firmly on the side of Akenhead J, to the effect that the payment under the Scheme did not constitute a new contract and that the period for issuing proceedings under the statute of limitations is to be judged by reference to the original contract.

In Ireland there is no legislation confining the period for instituting adjudication proceedings. The statute of limitations applies to arbitration by virtue of section 7(2) of the Arbitration Act 2010. There is no similar provision relating to adjudication, and amending legislation should perhaps be considered to provide for this lacuna if it is necessary. Andrew Bartlett, in a paper written for the Society of Construction Law, suggests:

> It cannot seriously be argued that the statutory provisions for adjudication, because of the phrase 'at any time' were intended to override the provisions of the Limitation Act. If an adjudicator finds that a claim under consideration is statute-barred, because the limitation period has expired without legal or arbitral proceedings being commenced, he or she will rightly dismiss it.[29]

Injunctions

The possibility of obtaining an injunction to prevent an adjudication proceeding should not be overlooked. It is unlikely, even in Ireland where constitutional issues will come into play, that an injunction would be obtained on the basis that the dispute is too complex for adjudication, or on the basis that the adjudicator could not possibly deal fairly with the dispute within the time prescribed. However, in the UK injunctions have been granted in circumstances where it would be oppressive to allow another set of adjudication proceedings to be issued where the claimant in those proceedings was in default in respect of previous adjudication determinations[30] and in circumstances where it was clear that the appointed adjudicator did not have jurisdiction.[31] However, the Outer House of the Court of Session in Scotland refused to grant an injunction in *T. Clarke (Scotland) Limited v Mmaxx Underfloor Heating Limited*[32] to prevent the defendant issuing a new set of proceedings where the defendant had already issued a number of vexatious proceedings by way of adjudication. It is to be noted, however, that the injunction was sought in circumstances where the new set of proceedings had not yet crystallised, and the court was not prepared to make the assumption that any new proceedings would be vexatious simply because the earlier proceedings could be so described.

Judicial review

The question of whether or not a party is entitled to seek judicial review of an adjudicator's decision has exercised the courts of a number of different states in Australia with varying results. The same issue will no doubt have to be addressed by the Irish courts in due course. The subject divides into four issues:

1   Is the decision of the adjudicator such as to fall within the ambit of judicial review?
2   If so, is the nature of the legislation such as to show an intention to exclude judicial review?

3   If it does show such an intention, is the legislature entitled to exclude the right to review such decisions from the jurisdiction of the courts?

4   If judicial review applies, is it in relation to both jurisdictional error and error on the face of the record?

To the extent that decisions may be made by the chair appointed by the Minister under the Irish Act in respect of the appointment of adjudicators to the panel, or appointment of adjudicators to individual cases, it is likely that such decisions, particularly the former, could be made subject to judicial review. The real issue is whether the adjudicator's decision itself is subject to such review. In *Brodyn Pty Limited (t/a Time Cost and Quality) v Davenport*[33] Judge Hodgson observed that: 'it is by no means clear that an adjudicator is a tribunal exercising governmental powers, to which the remedy in the nature of certiorari lies'.[34] However, it seems to be now well settled through the intervention of the High Court of Australia that an adjudicator's decision is potentially subject to certiorari or judicial review (the two being essentially the same in this context).[35]

Initially, judicial sentiment in New South Wales was that judicial review was available.[36] However, this view was found to be incorrect in *Brodyn* by the Court of Appeal. It found by way of necessary implication that judicial review did not apply, because the intention of the legislation was that it should not apply:

> The Act discloses a legislative intention to give an entitlement to progress payments and to provide a mechanism to ensure that disputes concerning the amount of such payments are resolved with the minimum of delay[37] . . . The availability of certiorari in such circumstances would not accord with the legislative intention disclosed in the Act that these provisional determinations be made and given effect to with minimum delay and minimum court involvement.[38]

However, as a result of the decision in *Kirk* and *Chase Oyster Bar*, Brodyn no longer applies. The *Kirk* decision made it clear that the legislature could not legislate against the entitlement to judicial review (certiorari) altogether. However, it did find that the legislature was entitled to legislate against judicial review in respect of error on the face of the record. As a result, the position in New South Wales is that a party is entitled to seek judicial review, but only in respect of jurisdictional error and not error on the face of the record.

In the State of Victoria the court found itself compelled to find that judicial review was available to the parties in all respects. This was because the Victorian Act specifically referred to the Constitutional Act of 1975 and it had to be inferred from this that the intention of the legislature was that judicial review would be applicable.

The position in New Zealand is far more straightforward. That judicial review applies to the adjudicator's decision has never really been in doubt.[39] In *Rees v Firth*[40] the Court of Appeal was asked to decide whether the entitlement to judicial review was confined to jurisdictional error. The court held:[41]

We are satisfied that the CCA (Construction Contract Act 2002) as a whole does not require that judicial review be limited to instances of what might be classified as jurisdictional error. In our view, to hold that the availability of judicial review is limited in that way invites unproductive and diversionary debate about whether a particular error is or is not 'jurisdictional'. The key point, we think, is that the statutory context is such that a person who does not accept the adjudicator's determination should litigate, arbitrate or mediate the underlying dispute rather than seeking relief by way of judicial review of the determination. Such relief will be available only rarely.

The court went on to say that the full gambit of judicial review reliefs would in principle be available, but that: 'it is unlikely that errors of fact by adjudicators will give rise to successful applications for judicial reviews'.[42]

It is thought likely that the Irish courts will decide that the parties are entitled to seek judicial review in principle in relation to an adjudicator's decision. The question then is whether error on the face of the record would be excluded from any such review as in the New South Wales example, or whether all aspects of judicial review are available, but that the court should use its discretion to discourage applications, as in New Zealand. The third alternative would be to find that judicial review does apply without limitation and not to discourage applications being made for such review.

It is thought that judicial review would not be available in England and Wales in respect of an adjudicator's decision made under the Scheme. This is because adjudication, even under the Scheme, arises out of an implied term of the contract, and judicial review is not available in respect of the decisions of a person appointed under a contract. In Scotland judicial review may be available in respect of an adjudication arising from a private contract. On the face of it, judicial review might offer a whole range of reliefs that would not be available through the Technology and Construction Court. If the full gambit of reliefs available through judicial review applied, an adjudicator's decision could be set aside if the adjudicator had made an error of law; if the adjudicator has been entirely unreasonable, within the meaning of the *Wednesbury* Rule; if the adjudicator had made findings of fact for which there was no evidence; if the adjudicator had adopted unfair procedures; or if the adjudicator had failed to give an intelligible determination. Considering, however, the manner in which judicial review is actually applied in respect of adjudication outside of England and Wales, it is questionable whether the reliefs that might be obtainable in England and Wales through judicial review, if it were available, would be any different to those currently available through the Technology and Construction Court.

Judicial review is available in Scotland, not just in relation to adjudications that take place under the Scheme but also in relation to those that take place under private contracts. One of the leading cases is that of *Diamond & Others v PJW Enterprises Limited*.[43] This was a judgment of the Second Division, Inner House, Court of Session by way of judicial review of a decision of an adjudicator of May 2001. It concerned a finding by an adjudicator that a professional firm of surveyors was responsible in negligence for damages payable to the respondent. The court was of the view that there was no legal basis to allow the adjudicator to come to such a decision. Nonetheless, it found at paragraph 42:

> In contrast with cases such as Watt v Lord Advocate (supra) or Joinery Plus Ltd v Laing Ltd (supra), where the decision-maker failed to understand the question that was remitted to him, in this case it is apparent that the adjudicator has at least answered the questions put to him in the notice of referral. Accordingly, I agree with the Lord Ordinary that the fact that the decision was erroneous on the question of professional negligence gives the petitioners no redress in proceedings of this kind.

At paragraph 40 the court had stated:

> The availability of judicial review as a remedy for an adjudicator's intra vires error of law would subvert the purpose of adjudication. If the courts were to interfere with a decision of an adjudicator on that ground, they would be adding a significant common law qualification to what is a statutory construct; they would be providing an opportunity for the kind of delay that the system is designed to prevent, and they would be providing a remedy which Parliament could have expressly provided but, it seems, chose not to.

Therefore, the court was firmly of the view that judicial review could only be availed of in relation to error of law to the extent that the error was one that was outside of the adjudicator's powers. However, if the TCC found that an adjudicator acted ultra vires, the TCC would also condemn the decision if the error had a significant effect upon it. The same result therefore is likely to be achieved whether judicial review is availed of or not.

The court was of the view that the role of judicial review was very limited. At paragraph 25 the court stated:

> The Lord Ordinary also suggested the possibility that the decision of an adjudicator should not be the subject of judicial review at all (at para [64]). Counsel for the respondents have not argued for that view, which I think goes too far. In my view, the court has a jurisdiction to review a decision of an adjudicator that proceeds on an erroneous exercise of jurisdiction; for example, where the adjudicator has exceeded his jurisdiction . . . or has failed to exercise it . . . or has exercised it fraudulently or in breach of natural justice. In each such case, the decision can be classified as being ultra vires.

These criteria, however, are applied equally by the TCC and are likely to be applied by any court being asked to enforce an adjudicator's decision. Essentially, it would appear that either on judicial review or in dealing with an application to enforce an adjudicator's decision, the courts will apply the same criteria and will not enforce a decision which is made ultra vires.

In that particular case, the adjudicator's decision was roundly criticised by the court in terms of being confused and lacking in reasoning. For instance, the adjudicator dealt with significant legal submissions made in the adjudication by simply stating that he had considered them in coming to his conclusion. No indication was given as to how he reconciled his conclusions with the authorities opened to him. Yet the court upheld the decision on judicial review. It would appear on the face of it that the court applied exactly the same principles in applying the concept of judicial review as the Technology and Construction Court would apply when dealing with an application to enforce a decision and that the bar was set no higher for the one than the other.

References

1 Building and Construction Industry Security of Payment Act 2004, Section 27(5)
2 *Auckland Water Proofing Limited v TPS Consulting Limited* Civ [2007]-4-5890
3 [2015] EWHC 412 (TCC); 159 ConLR 10, [2015] BLR 321, [2015] All ER (D) 01 (Mar)
4 Paragraph 59
5 New Zealand District Court Rules 2014, rules 20.86–20.88
6 *SW Global Resourcing Limited v Morris & Spottiswood Limited* [2012] CSOH 200, paragraph 34
7 [2013] EWHC 2954 (TCC); [2013] All ER (D) 38 (Oct)
8 [2005] EWHC 1086 (TCC) at paragraph 26; 101 ConLR 99, [2005] BLR 374, [2005] ArbLR 68, [2005] All ER (D) 277 (Jun)
9 *AWG Construction Services v Rockingham Motor Speedway* [2004] EWHC 888; 2004] EWHC 888 (TCC), [2004] All ER (D) 68 (Apr)
10 *Bouygues UK Ltd v Dahl-Jensen UK Ltd* [1999] All ER (D) 1281; (1999) 70 ConLR 41, [2000] BLR 49, [1999] Lexis Citation 3672,
11 *Rainford House Ltd (in administrative receivership) v Cadogan Ltd* [2001] BLR 416, [2001] All ER (D) 144 (Feb)
12 *Herschel Engineering Ltd v Breen Property Ltd,* unreported, 28 July 2000, Humphrey Lloyd J; (2000) 70 ConLR 1, [2000] BLR 272, [2000] Lexis Citation 3222, 16 Const LJ 366, [2000] All ER (D) 559
13 *Absolute Rentals Limited v Glencor Enterprises Limited,* unreported, 16 January 2000, Wilcox J; [2000] CILL 1637
14 [2008] EWHC 933 (TCC); 119 ConLR 130, [2008] 2 EGLR 7, [2008] 26 EG 118, [2008] All ER (D) 411 (Apr)
15 [2006] EWHC 741 (QB); [2006] All ER (D) 49 (Feb)
16 [2016] EWHC 241 (TCC); [2016] BLR 264, [2016] All ER (D) 165 (Feb)
17 [2004] NSWCA 394
18 [2010] NSWCA 190
19 [2010] EWHC 837 (TCC), 130 ConLR 90, [2010] BLR 452, [2011] Bus LR D42, [2010] All ER (D) 197 (Apr)
20 *Rupert Morgan Building Services (LLC) Limited v Gervis* [2004] 1 WLR 1867; [2003] EWCA Civ 1563, [2004] 1 All ER 529, 91 ConLR 81, [2003] NLJR 1761, [2004] BLR 18, (2003) Times, 26 November, [2003] All ER (D) 153 (Nov)
21 *Harris Calman Construction Company Limited v Ridgewood (Kensington) Limited* [2007] EWHC 2738 (TCC), [2008] BLR 132; [2008] Bus LR 636, [2007] All ER (D) 384 (Nov)

22 *Dalkia Energy and Technical Services Limited v Bell Group UK Limited* [2009] EWHC 73 (TCC), [2009] 122 ConLR 66; [2009] All ER (D) 273 (Feb)

23 Marcus Jacobs, Security of Payment in the Australian Building & Construction Industry, fifth edition, pages 428–429

24 [2007] EWHC 2507 (TCC), 115 ConLR 149, [2007] All ER (D) 444 (Oct)

25 [2014] EWCA Civ 93; 153 ConLR 26, [2014] 1 EGLR 97, [2014] All ER (D) 71 (Feb)

26 2002 SLT 781, 2001 SCLR 961, 2001 Scot (D) 21/7

27 [2013] EWHC 1322 (TCC); [2013] Bus LR 1199, [2013] NLJR 17, [2013] BLR 417, [2013] All ER (D) 296 (May)

28 [2009] EWHC 1906 (TCC); 125 ConLR 141, [2009] 3 EGLR 7, [2009] 41 EG 116, [2009] All ER (D) 29 (Aug)

29 The limits of adjudication: The impact of the European Convention on Human Rights, by Andrew Bartlett, Q.C. SCL Paper D175 dated December 2014

30 *Mentmore Towers Limited v Packman Lucas Limited* [2010] EWHC 457 (TCC); [2010] BLR 393, [2010] Bus LR D114, [2010] All ER (D) 236 (Oct)

31 *Twintec Limited v Volkerfitzpatrick Limited* (VFL) [2014] EWHC 10 (TCC); [2014] BLR 150, [2014] All ER (D) 177 (Jan)

32 [2014] CSOH 62; 2014 BLR 341; 2014 Scot (D) 12/3

33 [2004] 61 NSWLR 421

34 Paragraph 98

35 *Kirk v Industrial Relations Commission* [2010] HCA 1. While *Kirk* was not concerned with adjudication as such, it was applied to adjudication in *Chase Oyster Bar Pty Limited v Hamo Industries Pty Limited* [2010] NSW CA 190

36 *Musico v Davenport* [2003] NSWSC 1027, *Multiplex Construction Pty Limited v Luikens* [2003] NSWSC 1140 and *Tranagrid v Walter Construction Group* [2004] NSWSC 21

37 Hodgson JA, paragraph 51

38 Hodgson JA, paragraph 98

39 *Canam Construction (1955) Limited v James Christopher Lahatte & Yun Corporation Limited* Civ 2009 404 461

40 COA CA328/2011; [2011] NZCA 668

41 Paragraph 22

42 Paragraph 27

43 [2004] BLR 131, 2004 SC 430, 2004 SLT 545, 2004 Scot (D) 3/1

10 Right to suspend work

Introduction

Section 5 of the Act entitles the executing party to suspend work where a sum due under the contract (whether due under the statutory provisions or otherwise) is not paid. Section 7 entitles the executing party to suspend work where an amount due pursuant to the decision of an adjudicator is not paid. The former is regressive in comparison to standard existing provisions, and the latter may yet make a mockery of the Act.

Section 5: Right to suspend work for non-payment

5(1) *Where any amount due under a construction contract is not paid in full by the day on which the amount is due, the executing party may suspend work under the construction contract by giving notice in writing under subsection (2).*

(2) *Notice under this subsection shall specify the grounds on which it is intended to suspend work and shall be delivered to the other party—*

    (a) *not earlier than the day after the day on which the amount concerned is due, and*

    (b) *at least 7 days before the proposed suspension is to begin.*

(3) *Work may not be suspended under subsection (1)—*

    (a) *after payment by the other party of the amount due, or*

    (b) *after notice has been served by a party to the construction contract under section 6(2) in relation to a dispute relating to payment of the amount concerned.*

Section 5 provides that where an amount due under a contract is not paid in full by the due date, the executing party may suspend the work subject to notice. At least seven days' notice of the proposed suspension must be given. Subsection (3) provides that once payment is made the suspension can no longer be operated. Subsection 5(3)(b) provides that the suspension cannot proceed:

after notice has been served by a party to the construction contract under section 6(2) in relation to a dispute relating to payment of the amount concerned.

This provision is unique to the Irish Act and is difficult to justify. Suppose a certifier in accordance with subsection 4(3)(a)(i) specifies the amount to be paid in respect of an interim payment claim notice. If that amount is not paid within 30 days of the payment claim date, the executing party under this section 5 is entitled to suspend the works. However, that suspension, or threat of suspension, must be lifted if the party who was obliged to make the payment serves notice of adjudication in respect of the payment. Most standard forms of contract provide a right of suspension of works if any amount certified has not been paid. The Irish Act in this respect appears to be regressive in so far as it disallows a suspension in circumstances where adjudication is sought in respect of the payment. This provision is clearly open to abuse. It could allow the paying party, who is having difficulty in raising finance, to gain some extra time to enable it put the finance in place. In the meantime, the executing party has to carry on with the works and perhaps expose itself to an even greater financial risk. If the paying party is unable to raise the finance and becomes insolvent, the executing party will not only be out of pocket for the amount certified but also for the further work carried out in the period that it would otherwise have been entitled to suspend the works.

This begs the question as to whether clauses in standard contracts allowing a party to suspend work where a sum due under the contract has not been paid, must be read as incorporating a provision requiring the suspension to be lifted in the event of notice of adjudication being served. Presumably, they are to be so read unless the contract says otherwise and, if the contract says otherwise, that may be regarded as an attempt to limit or exclude the application of the Act and therefore be contrary to subsection 2(5) thereof.

As indicated, the modern tendency in legislation of this nature is to provide that where there is no response to the payment claim notice, the amount claimed, as opposed to the amount due under the contract, is then payable. Because the Irish Act adheres to the original philosophy of the UK Act of 1996 in this regard, the sum claimed through a payment claim notice is arguably not necessarily due and may be found in adjudication not to be due, notwithstanding that no response notice has been given. Where the modern tendency has been followed in other legislation, a claimant who serves a payment claim notice is not at risk by suspending the work if the respondent fails to respond and fails to pay the amount claimed. A party who suspends under the Irish Act is at risk of being found liable for damages if it transpires there was no sum properly due under the contract at the time the suspension occurred. This risk would appear to be reinforced by subsections 5(4) and 5(7), which appear to negate any entitlement to the executing party to suspend if it transpires that the suspension of work was unjustified. Subsection 5(4) provides:

Where work is suspended under subsection (1) and the ability of the execut-
ing party to complete work within a contractual time limit is affected by the
suspension of work, the period of suspension shall be disregarded for the
*purpose of computing the contractual time limit **unless the suspension of***
work is unjustified in the circumstances [emphasis added].

Some standard contracts provide that not only is the executing party entitled to
an extension of time for the period of the suspension but also for additional time
to allow for time wasted by reason of the demobilisation and remobilisation.
Whether or not such additional time can be claimed would depend upon the wording of the contract. It obviously does not arise under the Irish Act, but neither is
such a provision prohibited under the Irish Act.

Neither the Irish Act nor any of the other Acts expressly stipulate that the party
suspending will be entitled to recover all loss and expense incurred by reason of
the suspension. Some of the Australian Acts contain the following provision or
similar:

> If the claimant in exercising the right to suspend carrying out of construction
> work or the supply of related goods and services under construction con-
> tracts, incurs any loss or expenses as a result of the removal by the respondent
> from the contract of any part of the work or supply, the respondent is liable to
> pay the claimant the amount of the loss and expense.[1]

It is curious that the right to recover loss and expense is confined to that incurred:
'as a result of the removal by the respondent from the contract of any part of the
work'. This expression gives the impression that if work is suspended by virtue of
an entitlement under the Act, the respondent is entitled to remove that element of
the work from the contract and have it completed by another contractor.

Obviously, the intention of the legislature was that if the executing party suspended the works in accordance with its entitlement under the Irish Act, it would
be entitled to an extension of time for the period of the suspension. This requires
the person assessing the entitlement to the extension to have regard to the period
of suspension. Subsection 5(5), however, says that: 'the period of suspension shall
be disregarded' for the purpose of the calculation. Unfortunately, the legislature
would appear to have stated the exact opposite to what was intended. However,
if the legislation is in error in this regard, this presumably is a case where the
error must yield to the plain intention of the Oireachtas as required by section 5
of the Interpretation Act 2005, notwithstanding the recognised limitations of that
section.

Compare this with the relevant provision in the Isle of Man Act. This provides
at section 9(4) as follows:

> Any period during which performance is suspended in pursuance of the right
> conferred by this section shall be disregarded in computing for the purposes
> of any contractual time limit the time taken, by the party exercising the right

or by a third party, to complete any work directly or indirectly affected by the exercise of the right.

Where the contractual time limit is set by reference to a date rather than a period the date shall be adjusted accordingly.

Accordingly, under that legislation the suspension period is to be taken off the time actually engaged for the completion of the work for the purpose of assessing whether the executing party has met contractual time limitations. The Irish Act lacks that clarity.

Subsections 5(5) and 5(6) provide:

(5) *Where work is suspended under subsection (1) and the ability of a sub-contractor to complete work within a contractual time limit is affected by the suspension of work, the period of suspension shall be disregarded for the purpose of computing the contractual time limit.*

(6) *A period of suspension of work under subsection (1) shall also be disregarded for the purpose of computing the time taken to complete the work under another construction contract where—*

(a) *the construction contract the work under which is suspended is a subcontract,*

(b) *the other construction contract is also a subcontract and the other party to that other subcontract is the same as the other party to the subcontract the work under which is suspended, and*

(c) *the ability of the executing party under that other subcontract to complete work within a contractual time limit is affected by the suspension of work.*

Subsection 5(5) entitles any sub-contractor whose work is affected by such a suspension to an extension of time for the purpose of completing the sub-contract works, and subsection 5(6) entitles a different sub-contractor to the same principal to an extension of time in the event of its sub-contract works being affected by the suspension of other sub-contract works.

Notwithstanding that the right to suspend work is dealt with universally in such legislation, the Irish Act is the only legislation to contain these ancillary provisions, the rationale of which is not readily apparent. The intention of subsection 5(5) is at least clear. It is intended that a sub-contractor to the suspending party affected by a suspension shall be entitled to an extension of time. But why is the sub-contractor entitled to the entire period of the suspension? A suspension of two weeks may only have affected the sub-contractor's ability to complete its work by one week. The effect of subsection 5(6) appears to be to allow other sub-contractors affected by the suspension to an extension of time in respect of the full period of the suspension. The same criticism may be made in relation to it. Surely the entitlement to an extension of time should be measured against the extent to which the suspension affected the sub-contractor's ability to complete and not necessarily by reference to the duration of the suspension itself?

Subsection 5(7) provides as follows:

> *This section is without prejudice to the right of the other party to the con-*
> *struction contract under which work is suspended to claim for compensation*
> *or damages for any loss due to a suspension of work that is unjustified in the*
> *circumstances.*

Subsection 5(7) states that section 5 is without prejudice to the right of the other party, i.e. the paying party, to claim compensation for any loss due to a suspension that is unjustified. One might have expected this provision to be mutual, i.e. that it would state that it is also without prejudice to the right of the party suspending the works to seek compensation for the effects of the suspension where it is justified.

Section 7: Suspension of work for failure to comply with adjudicator's decision

(1) *Where any amount due pursuant to the decision of the adjudicator is not*
    paid in full before the end of the period of 7 days beginning with that on
    which the decision is made, the executing party may suspend work under
    the construction contract by giving notice in writing under subsection (2).

The wording here is identical to that at subsection 5(1), with the exception that the latter refers to any amount due under a construction contract as opposed to any amount due pursuant to the decision of the adjudicator.

(2) *Notice under this subsection shall specify the grounds on which it is in-*
    tended to suspend work and shall be delivered to the other party not later
    than 7 days before the proposed suspension is to begin.
(3) *Work may not be suspended under subsection (1)—*

    (a) *after payment by the other party of the amount due, or*
    (b) *after the decision of the adjudicator is referred to arbitration or*
        proceedings are otherwise initiated in relation to the decision.

These provisions mirror almost exactly what is in Section 5 relating to a failure to pay money due under the contract, and are open to the same criticism. It is submitted that it does not make sense that the right to suspend works ceases simply because the dispute, the subject matter of the adjudicator's decision, is referred to arbitration or litigation.

Throughout the adjudication world a party is entitled to suspend work if payment is not made on foot of an adjudicator's decision. That right does not cease because the dispute is referred to arbitration or litigation. Compare the Irish position with, for instance, that arising under section 26(4) of the Singapore Act:

Where a claimant has suspended the carrying out of construction work or the supply of goods or services under a contract in accordance with sub-section (1), he shall resume such work or supply within three days after being paid the adjudicated amount.

Under this legislation, therefore, which is typical, a party suspending is not obliged to resume work until payment is made, and even then it has three days grace to do so. Under the Irish Act, the suspending party is to lift the suspension immediately upon proceedings being initiated, irrespective of the fact that it has not been paid and, on a literal interpretation, irrespective of the fact that it may not yet have been served with the proceedings.

Subsections (4) and (5), which deal with the consequences of a suspension, are again almost identical to the equivalent subsections to Section 5 and are subject to the same criticisms made above in relation to it.

The wording of subsection 3(b) is unfortunate on two counts. First, it infers that what is being referred to arbitration or litigation is the adjudicator's decision rather than the dispute. Second, it would appear to provide that a suspension must be lifted where the party suspending initiates proceedings to enforce the adjudicator's decision.

The overall requirement for a 14-day period prior to suspension is far longer than would be required under other legislation. The position, for instance, under the Western Australian Act (Section 42) is that a contractor must give at least three days' notice.

References

1 Section 33(3), for example of the Queensland Act

11 Fees, costs and expenses

Introduction

The relevant provisions can be summarised as follows:

1   The parties must bear their own costs incurred in the adjudication. This is in keeping with most, but not all, legislation of other jurisdictions.
2   Adjudicators have a discretion in deciding who should pay their fees and expenses. This is the norm.
3   Adjudicators are entitled to be paid their fees in all circumstances. This is unique to Ireland.

Subsection 6(15): Parties' own costs

(15)   *Each party shall bear his or her own legal and other costs incurred in connection with the adjudication.*

This requirement follows the UK Act of 1996 and the general trend in such legislation internationally. As between the parties to the adjudication, costs are not recoverable. However, costs incurred by a party in an adjudication which would not have occurred but for the breach of contract or negligence of another party, may be recoverable in proceedings outside of the adjudication. In the recent decision of Akenhead J in *Board of Trustees of National Museums and Galleries on Merseyside v AEW Architects & Designers Limited*,[1] the court found that the designers were liable for the costs incurred by the plaintiff in defending adjudication proceedings brought by the building contractor against it. The court was satisfied that such expense was foreseeable in the sense of not being too remote.

With rare exceptions, international legislation providing for adjudication does not entitle the adjudicator to award party and party costs in any circumstances to one side or the other. However, the Singapore Act did cater for the precise point taken by Akenhead J in the Trustees of National Museums case. Section 30(4) of that Act provides:

A party to an adjudication shall bear all other costs and expenses incurred as a result of or in relation to the adjudication, but may include the whole or any part thereof in any claim for costs in any proceeding before a court or tribunal or in any other dispute resolution proceeding.

The reference to 'other costs' is a reference to costs other than the costs of the adjudication (i.e. the adjudicator's fees and expenses) and therefore includes the parties' own costs incurred in the adjudication.

The New Zealand Act is different in that it provides at section 56(1) that an adjudicator may determine that the costs and expenses incurred by one party will be met by the other party to the extent that those costs and expenses are unnecessarily incurred as a result of bad faith, or allegations or objections made without substantial merit. Importantly, section 56(2) goes on to say that absent such a determination, the parties to the adjudication must meet their own costs and expenses.

Similar to the New Zealand legislation, the Singapore legislation allows an adjudicator to award the costs incurred by one party in the adjudication against the other party in exceptional circumstances. In the case of the Singapore Act, this arises where an adjudicator is satisfied that a party to the adjudication incurred costs because of frivolous or vexatious conduct on the part of, or unfounded submissions by, another party.[2] The Western Australian Act is in similar terms to the Singapore Act.[3] These Acts also provide that subject to that discretion on the part of the adjudicator, the parties will each bear their own costs.

The Code of Practice at paragraph 38 provides that in Ireland, the parties are responsible for their own legal and other costs incurred in connection with the adjudication. The different wording of the legislation in the countries mentioned would not prevent, it is submitted, the application of the same logic as was applied in the *Trustees of National Museums* case.

Normally, where a party is awarded costs in court or arbitration proceedings, the costs are confined to the bare essentials, and a successful party will rarely recover its costs in full through taxation. Where, however, costs are awarded as damages, different principles apply. As long as the costs were reasonably incurred and the party to whom they are awarded met the usual obligation to mitigate, the cost should be recoverable in full.

In the UK it was common practice for institutions to include in their rules provisions entitling the adjudicator to decide not only who should bear the adjudicator's own fees and expenses but also the costs incurred by the parties themselves. The UK Act of 1996 was silent on the issue of party and party costs. This left parties free to agree through their contract how the costs would fall, and in *Bridgeway Construction Limited v Tolent Construction Limited*[4] it was held that this could include a provision whereby the party making the referral had to pay the costs of the other party. This provision has been modified by section 141 of the UK Act of 2009, as a result of which any agreement between the parties as to where the costs will fall is only valid if it is made in writing after the notice of intention to refer the dispute to adjudication has been given (this mirrors the very sensible provision of similar effect that had been contained at Section 30 of the Arbitration Act 1954, but was unfortunately omitted from the 2010 Arbitration Act). It would appear that in Ireland no agreement between the parties, whether it be through institutional rules or otherwise, entitling the adjudicator to determine who shall pay the costs incurred by the parties in the adjudication, will be valid. Subsection 6(15) is quite clear: the parties are each to bear their own costs. The only discretion vested in the adjudicator relates to the payment of the fees and expenses incurred by the adjudicator. Any attempt to alter this would be contrary to section 2(5) and in particular subsection (b) thereof.

Subsection 6(16): Costs of the adjudication

> (16)   *The parties shall pay the amount of the fees, costs and expenses of the adjudicator in accordance with the decision of the adjudicator.*

Compare this firstly with the UK Scheme as amended, which provides at paragraph 25 as follows:

> Subject to any contractual provision pursuant to section 108A(2) of the Act, the adjudicator may determine how the payment is to be apportioned and the parties are jointly and severally liable for any sum which remains outstanding following the making of any such determination.

The main difference between the two is that the Irish Act does not, through this provision at any rate, make the parties jointly and severally liable for the adjudicator's fees, whereas the English legislation does. As will be seen, subsection 6(17) of the Irish Act renders the parties jointly and severally liable for the adjudicators' fees if they resign, and section 6(18) makes the parties jointly and severally liable for those fees if they revoke the adjudicator's appointment. However, there appears to be a lacuna in the Irish legislation in that it does not provide for the parties being jointly and severally liable for the adjudicator's fees in the event of adjudicators rendering their decision. This is an anomaly, which will perhaps be addressed through the amendments one can anticipate to standard contracts by way of incorporating rules of adjudication.

It is to be noted that neither the Irish legislation nor the UK legislation indicates how adjudicators are to exercise their discretion in apportioning the costs of the adjudication as between the parties. In the UK adjudicators have followed the practice of judges and arbitrators arising from the basic concept that costs should follow the event. No doubt Irish adjudicators will follow suit.

The legislation in the non-European countries is quite diverse. Many, such as the NSW Act, the New Zealand Act, the Queensland Act, the South Australian Act and the Tasmanian Act provide that the parties will pay in equal shares unless the adjudicator decides otherwise. This infers perhaps a bias towards the adjudicator normally splitting their fees equally between the parties. The Malaysian Act expressly provides at subsection 18(1) that the costs of the adjudication are to follow the event, and the legislation of the Northern Territories and Western Australia both provide that the adjudicator's fees and expenses are to be paid in equal shares by the parties.

On the issue of joint and several liability, every one of these statutes expressly provides that the parties are jointly and severally liable for the payment of the adjudicator's fees.

Subsection 6(17): Resignation of adjudicator

(17) *An adjudicator may resign at any time on giving notice in writing to the parties to the dispute and the parties shall be jointly and severally liable for the payment of the reasonable fees, costs and expenses incurred by the adjudicator up to the date of resignation.*

It is difficult to understand the rationale behind this provision. If the adjudicator resigns, the parties are likely to have received little or no value from the adjudication process. Why in these circumstances should the adjudicator be entitled to any fee at all? It may indeed be appropriate to allow an adjudicator to resign in any circumstances and it may also be appropriate to exempt the adjudicator from any liability to the parties in those circumstances, but it is hardly appropriate to reward adjudicators by providing for payment of their fees up to the time of their resignation, given that the parties are likely to have incurred costs in the adjudication, which costs will be wasted, at least in part, by reason of the adjudicator's resignation.

This provision is open to abuse in that adjudicators who are not prepared to adhere to the time restraints imposed by the legislation may indicate to the parties that they will resign unless they are given further time to render their decision. Arguably, if adjudicators took such an approach without justification, they might be regarded as acting in bad faith and therefore not entitled to the immunity from liability afforded by subsection 6(14).

The UK Scheme provides at paragraph 9 that adjudicators will be entitled to be paid their reasonable fees where they decide that they do not have jurisdiction, or where the parties agree to revoke the appointment of the adjudicator. However, paragraph 11(2) expressly states that where the revocation of the appointment is due to the default or misconduct of the adjudicator, the parties have no liability for the adjudicator's fees and expenses. Given the extent to which the Irish legislation is based on that of the UK, the absence of a similar provision from the Irish Act is quite pointed.

The legislation across the remainder of the adjudication world is to the same effect. For example, the Singapore Act provides:

An adjudicator is not entitled to be paid, and shall not retain, any fee or expenses in relation to an adjudication application if he fails to make a determination on the application within the time allowed by section 17 or 19, as the case may be, otherwise than because the application is withdrawn or terminated or the dispute between the claimant and respondent is settled.[5]

This therefore is another provision peculiar to the Irish legislation. The basis upon which it was seen fit to depart from international practice is unclear.

The Irish Act does not provide for the basis on which the adjudicator's fees are to be assessed. Nor does it provide, as in the Arbitration Act 2010, for the adjudicator to assess their own fees and expenses. Therefore, the fees and expenses payable to the adjudicator are matters for agreement between the adjudicator and the parties. The Code of Practice provides at paragraphs 9 and 19 for the adjudicator providing the parties with their proposed terms of appointment including the basis for their fees, costs and expenses. It does not say what the consequences are if the parties do not agree to the terms or fees. Paragraph 36, however, does say that the adjudicator's fees, costs and expenses are to be reasonable in amount, having regard to the amount in dispute, the complexity of the dispute, the time spent by the adjudicator and other relevant circumstances.

Adjudicator's entitlement to be paid

The Irish Act does not provide for any circumstances in which adjudicators are not entitled to be paid their fees. Adjudicators are entitled to be paid fees if they resign, or their appointment is revoked, apparently irrespective of the reason for the resignation or revocation. But are adjudicators entitled to be paid fees if they provide a determination which is unenforceable? The legislation in non-European countries is clear on this. It states that provided adjudicators act in good faith, they are entitled to be paid their fees. The Irish legislation is silent on the point, as is the UK Act. However, the Court of Appeal has decided in relation to the UK legislation that an adjudicator is not entitled to be paid fees where the adjudicator's decision is unenforceable, whether that be as a result of lack of jurisdiction or otherwise.[6] The court concluded at paragraph 32:

> The purpose of the appointment was to produce an enforceable decision which, for the time being, would resolve the dispute. A decision which was unenforceable was of no value to the parties. They would have to start again on a fresh adjudication in order to achieve the enforceable decision which Mr Doherty (the adjudicator) had contracted to produce.

The court concluded therefore that the adjudicator was not entitled to be paid fees in respect of the adjudication. This has led in the UK to adjudicators insisting, wherever possible, upon the parties agreeing to pay their fees in these circumstances as a condition to accepting the appointment.[7]

It is perhaps ironic that adjudicators whose decisions may not be enforceable because of some important but understandable error as to jurisdiction on their part, would not be entitled to payment of their fees, whilst adjudicators who make a fundamental error and refuse to correct it, as in the *Bouygues* case (see page above), would be entitled to their fees. This is particularly so having regard to the fact that the courts have encouraged adjudicators to consider jurisdictional issues raised and, although they do not have the power to make a binding decision on jurisdiction, to proceed with the adjudication if they are of the view that the grounds relied upon to challenge jurisdiction are not valid.[8]

Adjudicators may attempt to include in their conditions a lien on their determinations in relation to their fees and expenses. Even, however, in the unlikely circumstances that both parties agree to this condition, it may be unenforceable as being contrary to the fundamental purpose of the legislation if the exercise of the lien would delay the determination beyond the date provided for by the legislation (*Mott MacDonald Limited v London and Regional Properties Limited,*[9] *Cubitt Building & Interiors Limited v Fleetglade Limited*[10]).

Having regard to the decision in *Systech International,* it obviously behoves adjudicators to provide in their terms and conditions that they will be paid their fees irrespective of whether the decision is found to be enforceable or not. Of course, it is not always feasible for an adjudicator to impose specific terms of contract as a condition to acceptance of the appointment. Where the adjudicator is appointed by the agreement of the parties, this will be usually possible. Where the appointment is by an appointing institution or, in Ireland, by the Chair of the Minister's Panel, this may not be possible. Very often a respondent will be delighted to avail of any opportunity to cause delay, or better still to abort the process. An adjudicator who will only accept the appointment if their terms and conditions are agreed, affords such a respondent that opportunity. Such adjudicators may find themselves without appointments if their insistence on imposing terms has this effect. The requirement at paragraph 19 of the Code of Practice that adjudicators shall provide the parties with their proposed terms of appointment may not always elicit agreement. The Code does not provide for this eventuality. Issues such as this may well be dealt with in the Code of Conduct, a draft of which has not yet been made available[11].

However, it is considered unlikely that the Irish Courts will follow the judgment of the Court of Appeal in *Systech International* given that the Irish legislation points strongly towards adjudicators being entitled to their fees in almost any circumstances.

Have adjudicators a lien on their decisions?

In the UK adjudicators do not have a lien over their decisions. Adjudicators cannot therefore refuse to furnish their decision until their fees are paid. This is contrary to the statutory regime operating in the non-European countries. Invariably, their legislation specifically provides for adjudicators being entitled not to communicate their decision until their fees are paid. There is no express provision dealing with this issue in the UK legislation, but there is in the legislation of the non-European countries. The Irish legislation is silent on the point. One might suspect that given there is no express provision in the Irish legislation entitling adjudicators to retain their decisions until their fees are paid, the position in Ireland will be the same as in the UK. However, this is by no means certain – the wording of the UK legislation is different to that of Ireland in relevant respects. Essentially, the courts have found in respect of the UK legislation that there is an overarching requirement for adjudicators to issue their decisions within the time prescribed by the legislation and that, whilst adjudicators may have a separate fee arrangement with the parties, this fee arrangement cannot override this requirement.[12] Moreover, the UK Scheme expressly provides at paragraph 19(3) that: 'as soon as possible after he has reached a decision, the adjudicator shall deliver a copy of that decision to each of the parties to the contract'. There is no such provision in the Irish Act or in the Code of Practice. The adjudicator is required by the legislation to reach a decision within the 28 days (or such extended period as may have been agreed), but the Act is silent as to when the adjudicator's decision is to be communicated.

Although the Irish Act does not provide any express limitation on the period adjudicators have to communicate their decision as opposed to reaching that decision, it is to be noted that under subsection 7(1) a right of suspension arises if any sum payable under the decision is not paid within seven days of the decision being made. This must give rise to an inference that the decision is to be communicated promptly.

Whether or not, therefore, the Irish legislature intended the adjudicator to have a lien is open to debate, given that the legislation has not followed that of the other European countries or that of the non-European countries.

If an adjudicator does have a lien in respect of the determination, or is entitled to create such a lien through terms and conditions of acceptance of the appointment, adjudicators must be very careful as to how they exercise that entitlement. In *Mott MacDonald Limited v London & Regional Properties Limited*[13] the adjudicator was found to have breached the obligation of impartiality by imposing a condition whereby the referring party was obliged to pay all of the adjudicator's fees for the release of his determination. An adjudicator/arbitrator must be careful to treat the parties equally in every respect.

Subsection 6(18): Revocation of the adjudicator's appointment

(18)   *The parties to a dispute may at any time agree to revoke the appointment of the adjudicator and the parties shall be jointly and severally liable for the payment of the reasonable fees, costs and expenses, incurred by the adjudicator up to the date of the revocation.*

Again, it is difficult to understand why, in all circumstances, the parties would be obliged to discharge adjudicators' fees and expenses where their appointment is revoked. If appointments are revoked because of the incompetence or misconduct of adjudicators, or by reason of adjudicators not abiding by the Code of Practice, one would expect that the adjudicators should not be entitled to be paid. However, under this provision they are entitled to be paid their fees, costs and expenses irrespective of the reason for the revocation of the appointment.

This is another provision that is unique to the Irish Act. The legislation elsewhere invariably provides that adjudicators are not entitled to be paid fees and expenses if they fail to provide their decisions within the time prescribed, subject to specific exceptions.

Subsection 6(14): Adjudicator exempt from liability

(14) *The adjudicator is not liable for anything done or omitted in the discharge or purported discharge of his or her functions as adjudicator unless the act or omission is in bad faith, and any employee or agent of the adjudicator is similarly protected from liability.*

This is not an unusual provision and it would be surprising, having regard to the provisions entitling adjudicators to their fees in any event if this provision was otherwise. It is to be noted that there is no express provision to the effect that adjudicators are not entitled to their fees and expenses even if they act in bad faith. Most of the legislation providing for adjudication contains similar provisions. The Singapore Act, for instance, states: 'No suit or other legal proceedings shall lie against an adjudicator with respect to anything done or omitted to be done in good faith in the discharge or purported discharge of its functions or duties under this Act'.

An interesting comparison can be made with the equivalent provision in the Arbitration Act 2010, which at section 22 provides: 'An arbitrator shall not be liable in any proceedings for anything done or omitted in the discharge or purported discharge of his or her functions'.

It would appear that arbitrators are exempt from liability even where they act in bad faith, but an adjudicator is not. Given that an arbitrator's decision is final, whereas an adjudicator's is not, one would think that it would be just as important, if not more important, that an arbitrator would be accountable for acting in bad faith.

References

1 [2013] EWHC 3025; [2013] EWHC 3025 (TCC), [2014] 1 Costs LO 39
2 Building & Construction Industry Security of Payment Act 2009, Section 30(3)
3 Construction Contracts Act 2004, Section 34(2)
4 [2000] CILL 1662
5 Building & Construction Industry Security of Payment Act 2009, Section 31(2)
6 *Systech International Limited v PC Harrington Contractors Limited* [2013] 2 All ER 69; [2012] EWCA Civ 1371, [2013] 1 All ER (Comm) 1074, [2013] Bus LR 970, 145 ConLR 1,[2013] 1 EGLR 9, [2013] BLR 1, [2013] 03 Estates Gazette 88, (2013) Times, 01 January, [2012] All ER (D) 283 (Oct)
7 *Linnett v Halliwells LLP* [2009] EWHC 319 (TCC) at [76]-[79], (2009) 123 ConLR 104
8 *Enterprise Managed Service Limited v McFadden Utilities* [2009] EWHC 3222 [2009] EWHC 3222 (TCC), [2011] 1 BCLC 414, [2010] BLR 89, [2010] All ER (D) 126 (Apr)
9 [2007] EWHC 1055 (TCC); 113 ConLR 33, [2007] All ER (D) 431 (May)

10 [2006] EWHC 3413; 110 ConLR 36, [2007] All ER (D) 268 (Jan)
11 *When applications were invited for membership of the Minister's panel, it was indi-cated that members of the panel would have to sign up to a code of conduct yet to be published*
12 *St. Andrews Bay Development Limited v HBG Management Limited* [2003] SLT 740; 2003 SCLR 526, 2003 Scot (D) 19/4; *Cubitt Building & Interiors Limited v Fleetglade Limited* [2006] EWHC 3413 (TCC); [2006] 110 Con LR 36; [2007] All ER (D) 268 (Jan)
13 [2007] EWHC 1055 (TCC); 113 ConLR 33, [2007] All ER (D) 431 (May)

12 Code of Practice

Introduction

Section 9 of the Act entitles the Minister to prepare and publish a Code of Practice governing the conduct of adjudications. Stakeholders such as The Construction Industry Federation, Engineers Ireland and other representative bodies of the relevant professions were invited to comment upon a number of drafts of the Code of Practice and following such comments the Code of Practice was passed into law on the 5th July 2016 but quickly revoked and replaced by a second Statutory Instrument on the 25th July 2016. These Statutory Instruments contained a number of provisions that were not included in earlier drafts. The stakeholders were not given an opportunity of commenting upon some of the provisions that are considered least desirable by the author such as the new elements contained in paragraphs 8 and 11 as set out below.

The Code of Practice comprises thirty-nine numbered paragraphs which are grouped under the following headings:-

Paragraph	1	Definitions
Paragraphs	2–3	General
Paragraphs	4–5	Preliminary
Paragraphs	6–7	Prospective Adjudicator responsibilities to the parties to a payment dispute.
Paragraphs	8–12	The Appointment of an Adjudicator – by agreement of the parties.
Paragraphs	13–20	The appointment of an Adjudicator – by the Chairperson.
Paragraphs	21–22	Referral of a payment dispute to an Adjudicator.
Paragraphs	23–38	Adjudication of a payment dispute – Procedures and Decision
Paragraph	39	Reporting on the Conduct of Adjudication Cases.

It is unusual to publish a Code of Practice to supplement primary legislation. It is quite normal to provide for the Minister concerned having a power to introduce from time to time regulations to supplement such legislation and to some extent the Code of Practice can be likened to regulations. The draftsman of the Act had primary regard for the legislation in the UK when preparing the legislation and it may be that the Code of Practice was seen as being akin to the Scheme published in that jurisdiction.

It is proposed to comment paragraph by paragraph upon the contents of the Code of Practice with the exception of the definitions which do not require such comment.

General

2   *The procedures set out in this Code of Practice shall apply to each individual payment dispute arising under the Act. In accordance with section 6(9) of the Act, an Adjudicator may deal at the same time with several payment disputes arising under the same construction contract or related construction contracts.*

3   *No liability whatsoever shall extend to the Minister, Chairperson or to the Department of Jobs, Enterprise and Innovation in respect of this Code of Practice or for any loss that arises from the operation of this Code of Practice. The Minister reserves the right to make changes to this Code of Practice.*

The last sentence of paragraph 3 is superfluous. Either Ministers have the power to make changes to the Code of Practice from time to time or they do not. If not, the Minister cannot create such an entitlement by merely stating it in the Code of Practice. In fact under sub-section 22(3) of the Interpretation Act 2005 the power conferred on a Minister to create a statutory instrument includes the power to amend or revoke that statutory instrument. As it happens the Minister has already exercised this power by replacing the statutory instrument of the 5th July 2016 with that of the 25th July 2016

Preliminary

4   *A party to the construction contract (known as 'the Referring Party') commences adjudication pursuant to section 6(2) of the Act by serving a written Notice of Intention on the other party or parties to the construction contract (known as the 'Responding Party/Parties') under which an individual payment dispute arises.*

In practice advisers to contractors and subcontractors in the UK often recommend delivering a draft Notice of Adjudication (Notice in Intention in Irish terminology) or even a draft Notice to Refer and the accompanying documents on the other party a month or so in advance of the actual commencement of adjudication proceedings. It is not always considered appropriate, or in the referring party's interests, to refer a complicated dispute immediately to adjudication. If the claim has been delivered with the same detail to the respondent in advance of the formal service, the chances of problems arising by reference to natural justice are diminished. Furthermore the parties through this process are given the opportunity of resolving the issue before incurring the expense of an adjudication. Very often the delivery of the draft notice of adjudication or referral will be accompanied by an invitation to meet to discuss the matter with a view to resolving it. The names of possible adjudicators might also be put forward in the hope that, if the dispute cannot be resolved, it will at least be dealt with by an adjudicator trusted by both parties. While the ambush tactic in the UK has not gone away, it is by no means as prevalent as it once was. This procedure is the very opposite to an ambush.

The tactic of ambush was discussed by Akenhead J In the case of *Bovis Lend Lease Limited v Trustees of the London Clinic*[1] in which it was stated:

[50]   *Various other matters of principle are also agreed (properly in my view) by the parties:*

(a)   *The fact that a dispute is complex or involves consideration of large volumes of material does not necessarily mean that any decision reached within an adjudication is procedurally unfair (see CIB Properties Ltd v Birse Construction [2005] 1 WLR 2252, [2005] BLR 173).*

(b)   *The mere fact that there has been an "ambush" by the claiming party in an adjudication does not in itself amount to procedural unfairness (see London & Amsterdam Properties Ltd v Waterman Partnership Ltd [2003] EWHC 3059 (TCC) at para 179, 94 Con-LR 154, [2004] BLR 179).*

[51]   *With regard to the question of "ambush", the statutory framework of the Housing Grants Construction and Regeneration Act 1996 is one which enables a party to a construction contract to refer anything, which might be classified as a dispute to adjudication, in the ordinary course of events for a decision to be provided by the adjudicator within 28 days of the reference. Therefore, the threshold to a reference to adjudication is simply and only that there is a crystallised dispute. Thus, if a dispute has arisen by 23 December in a given year, the referring party may refer that dispute to adjudication on 24 December. That might give rise to an assertion that there has been an "ambush" because the defending party may well have insufficient time, given the Christmas break common in the construction industry, to prepare its defence. It is not uncommon, similarly, for claiming parties to refer matters to adjudication during the summer holidays when it is known that key personnel of the defending party are away. Again, this might be said to be an "ambush". However, for better or for worse, Parliament does not expressly give an adjudicator the power to extend the 28 days by reason of that fact. However, there is a sensible school of thought which suggests that in those circumstances an adjudicator*

can in effect decline to accept the appointment on the grounds that justice cannot be done or the adjudicator can simply say to the claiming party words to the effect: "Unless you agree to an extension of time I will not be able to produce my decision within 28 days." Indeed, that is commonly what adjudicators will do and it is a very rare case when the claiming party does not accede to some extension of time accordingly.

It is not just the twenty-eight day period that gives rise to a potential ambush, it is also in the UK the fact that only a period of seven days is allowed between service of the notice of adjudication and service of the referral to arbitration. The respondent may therefore find itself fully engaged in adjudication without any opportunity to fully consider its position. Crucially, in Ireland a significantly longer period will elapse between the service of the Notice of Intention to refer and the referral (see commentary on paragraph 14 of the code below).

5   *A Notice of Intention shall include:*

(i)   *the name, address and contact details of each party to the construction contract;*

(ii)   *relevant details of the payment dispute to include the amount in dispute (even if the amount is zero), the nature of the payment dispute, and the site address;*

(iii)   *a copy of the relevant payment claim notice, and any response to that payment claim notice as provided for in section 4 of the Act; and*

(iv)   *relevant details to identify the construction contract and any supporting information that may assist an Adjudicator in understanding the nature of the payment dispute. Where a written construction contract exists, this must be attached.*

Neither the Act nor the code indicate the consequences of non-compliance with the statutory requirements. Does failure for instance to include any of the items (i) to (iv) as listed above invalidate the notice? The case law in the UK would suggest that this depends upon whether or not the omission has a significant consequence. The same case law would suggest that even if there is a defect in the notice, if the responding party fails to make its objection clear to the adjudicator, it will not be entitled to do so by way of challenging the adjudicator's decision.

Given that the Act applies to oral contracts as well as contracts in writing, item 5(iv) may give rise to difficult issues of contract interpretation. When and how the contract was formed may not be readily ascertained. It is unlikely that incorrect information in this respect will invalidate the notice. The process is clearly intended to be availed of without legal assistance and it is therefore unlikely to be interpreted in such a way that legal advice might be essential at the outset.

The draft of the code immediately preceding the final version also required copies of relevant payment claim notices and responses to payment claim notices but included the words "*if any*". It is to be noted that these words have been omitted from the final version. This is consistent with a suspicion the stakeholders hold that initially the Department intended payment disputes to be confined to those arising from a payment claim notice. Initial drafts prepared by the Department appeared to indicate that this was the case. It may be that by the deletion of the words "*if any*" the Minister, through the Code of Practice, is seeking to confine disputes to those arising from payment claim notices. This of course cannot be done if such confinement is in conflict with the Act itself. The Act clearly entitles a party to refer "any" payment dispute to adjudication.

Item 5(ii) may indeed have the opposite effect. If a payment dispute can relate to a claim where the amount involved is zero, this would suggest that a dispute relating to payment may include claims seeking a decision on liability without necessarily addressing the quantum. For instance, it might be argued that a claim seeking an extension of time for completion of the contract would relate to payment but, if no sum is actually sought in the application, the amount in dispute would be zero. The justification for the inclusion of the words "*(even if the amount is zero)*" is probably the inclusion of those words at section 4(2)(a) of the Act. That anomaly is discussed above at page 39.

It is to be noted that a copy of the written construction contract, if it exists, must be attached to the Notice of Intention. A further copy must be provided to the respondent with the application to the chairperson referred to at paragraph 15 of the code and a further copy must again be provided with the referral itself (paragraph 22). It should be borne in mind that the full contract must be provided and not just the relevant extracts. The contract might include thousands of pages and dozens of drawings. This would appear to be an unnecessary burden imposed upon the parties, not to mention the environment.

Prospective Adjudicator's responsibilities to the parties to a payment dispute

6   *A prospective Adjudicator should only accept an appointment to a payment dispute under the Act if he/she:-*

(i)   *is able to give the adjudication the time and attention which the parties to the payment dispute are reasonably entitled to expect;*

(ii)   *believes that he/she is competent to determine the issues in dispute; and*

(iii)   *is satisfied that no conflict of interest exists between him/her and the parties subject to paragraphs 11 and 20 of this Code of Practice.*

That the adjudicator should have sufficient availability is perhaps obvious. It is of course difficult for adjudicators to estimate at the outset the time required. Busy adjudicators in the UK tend to have a number of adjudications running at the same time. The obligation to consider one's own competence in terms of the issues in dispute is a necessary one. It may not be appropriate for instance for a lawyer to accept an appointment where the dispute involves highly technical engineering or architectural concepts.

It might have been appropriate to also require adjudicators specifically to consider whether it is possible to fairly issue a decision within twenty-eight days. The Irish courts may be less tolerant of adjudicators making rash assumptions in this regard. In *O'Brien's Irish Sandwich Bars Limited v Companies Acts*[2] the High Court decided that it could not deal with an application for examinership in circumstances where twelve hearings might be required to allow different landlords have a fair opportunity to be heard as to whether the applicant company should be released from the terms of its leases. This was because the court simply would not be able to guarantee fair procedures within the time allowed by the legislation. Accordingly the company was put into liquidation. Subject to the view the courts may take of the fact that the adjudicator's decision is only temporary, the courts would be unlikely to accept lesser standards in adjudication.

> 7. *A prospective Adjudicator shall not contact any party to a payment dispute under the Act in order to solicit appointment as an Adjudicator to that dispute.*

This is a modest impediment upon adjudicators. There is no impediment as to the extent to which adjudicators may otherwise solicit appointments through advertising or otherwise. The prohibition against soliciting only arises once a payment dispute comes into existence.

The Appointment of an Adjudicator – by agreement of the parties

> 8   *The parties to the construction contract may, within five days beginning with the day on which the Notice of Intention is served, agree to appoint an Adjudicator of their own choice and he / she may be a person referred to in the construction contact to perform that role, a person from the Panel or he/she may be another suitably qualified person.*

This provision has sent shockwaves through the industry in so far as it infers that parties are entitled to agree an adjudicator through their contracts. This entitlement was not referred to in any previous drafts and a reading of the Act on its own suggested that this was not possible. The Act at sub-section 6(3) allows the parties to agree an adjudicator of their own choice within the five day period beginning with the day on which Notice of Intention is served. Failing such agreement sub-section 6(4) requires the parties to apply to the chairperson to make an appointment from the panel of adjudicators. Depending on how these sub-sections are interpreted, it may be that the Act prohibits the appointment of an adjudicator through their written contracts. The Code of Practice cannot overrule the Act and any such appointment would therefore be unlawful. However, unless and until a decision is made by the courts on this issue, no doubt employers and main contractors will seek to include adjudicators of their choosing in their contracts.

A literal interpretation of this paragraph would suggest that the parties may nominate or appoint an individual as their adjudicator in their contract but that person will not be appointed unless the parties within the five day period so agree. Such a literal interpretation however does not make much sense. Such an appointment through the contract would be meaningless in those circumstances. On the other hand the wording of paragraphs 9 and 39 of the Code of Practice seems to reinforce the view that appointments can only be made through the procedures provided for in sub-section 6(3) and 6(4) of the Act.

It must be borne in mind that the Minister through the statutory instrument constituting the Code of Practice cannot make laws. The Oireachtas is the only body entrusted with the making of laws as such. A statutory instrument must remain within the framework of the primary legislation. The test for deciding whether secondary legislation of this nature goes beyond its remit was stated by O'Higgins C. J. in *Cityview Press v An Chomhairle Oiliuna*[3]:

> In the view of this Court, the test is whether that which is challenged as an unauthorised delegation of parliamentary power is more than a mere giving effect to principles and policies which are contained in the statute itself. If it be, then it is not authorised; for such would constitute a purported exercise of legislative power by an authority which is not permitted to do so under the Constitution. On the other hand, if it be within the permitted limits – if the law is laid down in the statute and details only are filled in or completed by the designated Minister or subordinate body – there is no unauthorised delegation of legislative power.

If the Act at sub-sections 6(3) and 6(4) is clear as to the only manner in which an adjudicator can be appointed, and it is submitted it is, the Code of Practice cannot be interpreted as adding an additional or conflicting mechanism for the appointment.

In the author's opinion it would be most unfortunate if the parties were entitled to appoint adjudicators through their contracts. In practice this gives the parties who prepares the contract the choice of adjudicator because contractors or subcontractors, as the case may be, are usually offered terms of contract on a take it or leave it basis (see page 64 above).

9   *A person who is requested to accept an appointment as Adjudicator following an agreement by the parties to the construction contract in accordance with section 6(3) of the Act shall, within two days of such a request and prior to accepting the appointment, write to the parties to ask them to disclose any information indicating any potential conflict of interest that may arise from the person's appointment as Adjudicator. He/she shall draw the attention of the parties to the provisions of paragraph 32 of this Code of Practice. The prospective Adjudicator shall, at the same time provide the parties with his/her proposed terms and conditions of appointment, including the basis for his/her fees, costs and expenses.*

The reference in the first sentence to section 6(3) suggests that the only mode of appointing an adjudicator by agreement is through that provision and this supports an interpretation of the legislation to the effect that any appointment through the contract will not apply unless it is confirmed under section 6(3). The first sentence of paragraph 13 gives rise to a similar inference.

The Code of Practice published by Statutory Instrument on the 5th July 2016 put the onus on the parties to make disclosure of any perceived conflict of interest on the part of the appointee. The Code of Practice now puts the onus on the appointee. It might have been appropriate to require both the appointee and the parties to make such disclosure. It is submitted that it is appropriate for prospective adjudicators to declare not only potential conflicts of interest but also such professional relationships as might exist, such as the fact that the adjudicator and one of the lawyers involved sit on a committee together or regularly play golf together. Such relationships do not of themselves give rise to a conflict of interest but, if not declared at the outset can give rise to suspicions if later discovered by a party.

Paragraph 32 of the code is obviously considered of great importance by the draftsman. The obligation to draw the parties attention to that provision is emphasised here, at paragraph 19 and also at paragraph 26.

The inclusion of the words *"(t)he basis for his/her fees, costs and expenses"* in the final sentence is somewhat confusing. There was no obligation to provide such a basis in previous drafts and it is difficult to understand what exactly is meant by this expression. It may be that the adjudicator is obliged to justify the fees being proposed with reference to the requirements of paragraph 36 which requires that the fees, costs and expenses shall be reasonable in amount having regard to the amount in dispute, the complexity of the dispute and other relevant circumstances.

10   *Each party shall within three days of the communication from the prospective Adjudicator decide if the appointment of the prospective Adjudicator is to proceed and inform the prospective Adjudicator in writing of their decision.*

This paragraph does not provide for the consequences of one of the parties failing to respond within the three days. There is no default position. It is clear that if one party rejects the appointment, that is the end of the matter but if one party simply does not respond, is it deemed to have accepted the terms and conditions proposed by the adjudicator or to have rejected them?

This raises an issue as to whether an adjudicator is entitled to seek a deposit in relation to his fees and expenses. On the face of it, there is no reason why not. On the other hand, if only one party agrees to pay the required deposit and the other party does not, this can raise the spectre of apparent bias (*K/S Norjarl v Hyundai Heavy Industries*[4]).

11 *If a potential conflict of interest is disclosed by any party, the prospective Adjudicator may subject to the consent of all the parties, and on satisfying any professional and/or ethical concerns he/she may have, accept the appointment.*

This provision was not included in any earlier draft and therefore the stakeholders had no opportunity to comment upon it. It is not of particular concern in the context of an adjudicator being appointed by the agreement of the parties. The statutory instrument of the 5th July 2016 contained a similar provision in respect of an adjudicator appointed by the chairperson of the panel. The main purpose to revoking that statutory instrument was the amendment of that provision. This provision allows a reluctant respondent (not a rare phenomenon) an opportunity to create delay even in the context of the appointment being made by the agreement of the parties. To buy time, the respondent might agree the name of an adjudicator, and might even agree the adjudicator's terms and conditions, but then raise a potential conflict of interest on the part of the adjudicator. It is to be noted that there is no time limit on raising the potential conflict of interest. There is certainly an inference that this should be done within the three days but the code does not specifically say so. It is quite possible that a genuine conflict of interest would be identified outside of the three days because, for example, the person who would have known of the conflict was on holidays at the time the appointment was agreed. One would assume therefore that the intention of paragraph 11 is that it would apply irrespective of when the potential conflict is disclosed.

It is to be noted that an actual conflict is not required. Potential conflict is sufficient. No doubt reluctant respondents will make great play of this distinction.

It is also to be noted that even if the parties are agreeable to the adjudicator proceeding despite a potential conflict of interest, the adjudicator is obliged to examine his own conscience in that regard before accepting the appointment.

12 *If the appointment of the prospective Adjudicator is to proceed, the prospective Adjudicator shall write to each party to accept the appointment and the date of the letter of acceptance sent to the parties shall be deemed to be the date on which the appointment of the Adjudicator is made for the purposes of section 6(5) (a) of the Act. Such acceptance, anonymised in terms of the details of the parties to the dispute, shall be notified by the Adjudicator to the Construction Contracts Adjudication Service of the Department of Jobs, Enterprise and Innovation for the purpose of compiling statistical information relating to the Act.*

It is necessary that the exact date of the appointment of the adjudicator is known because the referral must be made within seven days of the appointment beginning with the day on which the appointment is made (section 6(5)(a)). The final sentence is to be read in conjunction with paragraph 39. It is important that statistical information is available and monitored for the purpose of future reviews of the legislation. The statistics available in the UK relate only to appointments made by authorised appointing bodies, as opposed to appointments made by the agreement of the parties. This provision should ensure that the statistical information available in Ireland will be complete.

The Appointment of an Adjudicator – by the Chairperson

13   *Failing agreement by the parties to select an Adjudicator in accordance with section 6(3) of the Act, a party to the construction contract may apply to the Chairperson to seek the appointment of an Adjudicator from the Panel in accordance with section 6(4) of the Act. Relevant contact details are available on the website of the Department of Jobs, Enterprise and Innovation at www.djei.ie.*

14   *If an application is to be made under section 6(4) of the Act to the Chairperson, it shall be made not earlier than five days from and including the day on which the Notice of Intention was served.*

Paragraph 13 does not require comment. It merely repeats what is in the Act.

The Act itself does not prevent an application being made to the chairperson immediately upon service of the Notice of Intention to Refer. Paragraph 14 does do so and gives the responding party five days respite. Section 108 of the 1996 UK Act requires that the referral to the adjudicator be made within seven days of the notice of adjudication. This is an extremely tight timeframe and has the potential to be most unfair to the respondent. In Ireland there is likely to be normally a delay of about twelve to nineteen days between the Notice of Intention to refer and the referral where the parties do not agree upon an adjudicator, because five days must lapse under this paragraph before an application is made to the chairperson, the chairperson is likely to make the appointment seven days later and the referral is then to be served within a further seven days of the appointment.

There is no time limit for making the application to the Chairperson after the five days has expired. There is a period therefore in which the parties may negotiate, if they so wish, before the application is made. It is not clear whether an appointment made by agreement of the parties outside of the five day period would be valid. There is no similar provision in the UK legislation but time limits of this nature in that legislation have been applied rigidly by the courts on the basis that Parliament clearly intended such time limits to be mandatory. Therefore if the referral is not made within seven days of the notice of adjudication, the process has to start again. Presumably the Irish courts would take a similar attitude as regards the period of twenty-eight days for making a decision, and similar time limits but it is unlikely that the court would be so rigid in other respects given that aspects of the process, such as the time for making application to the Chairperson are open ended. It is thought therefore that the purpose of the five days is to allow a mandatory period for the parties to agree an adjudicator prior to which either party can apply to the Minister. It is not to prevent the parties agreeing outside of the five days to make an appointment of their own. However, as with so many other provisions of the Act, this is a point that can be argued either way. Taken literally sub-section 6(4) requires the appointment to be made by the Chairperson unless the parties have agreed to make an appointment within the five days. That such a literal interpretation should not be adopted is perhaps supported by paragraph 10 of the code. Presumably the three day period referred to in that paragraph is not expected to occur within the five days allowed by sub-section 6(3). Therefore the agreement reached within the five days is, at best, a conditional agreement subject to the terms and conditions of the adjudicator, and subject to there being no conflict of interest. It may be that the actual agreement to appoint only occurs when both parties respond positively within that three day period.

15   *An application to the Chairperson to appoint an Adjudicator from the Panel to a payment dispute shall be in writing and submitted to the Chairperson in accordance with the application procedures set out by the Construction Contracts Adjudication Service of the Department of Jobs Enterprise and Innovation from time to time. Such application, shall be copied by the applicant to the other party/parties to the payment dispute at the same time and shall include:*

(i)   *the name, address and contact details of each party to the construction contract;*

(ii)   *relevant details of the payment dispute to include the amount in dispute (even if the amount is zero), the nature of the payment dispute, and the site address;*

(iii)   *a copy of the Notice of Intention including any accompanying documents attached to that Notice;*

(iv)   *the date as to when the Notice of Intention was served on the Responding Party/Parties and how this was done; and*

(v)   *relevant details to identify the construction contract and any supporting information that may assist an Adjudicator in understanding the nature of the payment dispute. Where a written construction contract exists, this must be attached.*

The wording of items (i), (ii) and (v) corresponds verbatim to items (i), (ii) and (iv) of what is to be included under paragraph 5 in the Notice of Intention. In so far as item (iii) requires that a copy of the Notice of Intention including any accompanying documents is to be provided, these items appear to be superfluous. All that is required are items (iii) and (iv). It is of course important that the Notice of Intention be served in accordance with the contract or, in default of any provision in the contract, in accordance with section 10 of the Act.

The requirement that a copy of the request for the appointment be delivered to the responding party is laudable given the content of paragraph 15(v). This allows a party to perhaps steer the chairperson towards making an appointment of a specific person or from a narrow category of persons. It is appropriate that the responding party would have an opportunity to respond to any such invitation prior to the appointment being made, and it is unlikely that an appointment will be made until late into the seven day period normally allowed to the chairperson for making the appointment for this reason.

Controversial information which may be provided by the referring party to the chairperson may take the following form:-

1   The referring party may depict the nature of the dispute in such a way as to indicate or infer that an adjudicator having a particular skill set is required. It could be that the aspect requiring such skills is only a minor element and that other skills might be more appropriate to deal with the main issues and that the referring party is mainly seeking to achieve the appointment of a particular person.

2   The referring party might advise the chairperson of the names put forward by each of the parties to the other for approval. Some institutions, when appointing arbitrators, have a policy of not selecting anyone who has been rejected by either party. It remains to be seen whether the chairperson will adopt such a policy.

3   The chairperson may be advised that a particular person has already adjudicated upon related disputes arising between the same parties out of the same contract. The referring party may suggest that it would be appropriate to appoint that party because of their previously acquired knowledge of complicated issues relating to the parties or the contract. The referring party is not likely to trouble the chairperson with such information if it fared badly in the earlier adjudication.

The code does not expressly entitle the responding party to comment upon the request or to independently supply information it considers relevant. It is to be inferred however that the responding party would have such a right and that the responding party would be obliged to provide the referring party with a copy of any such communication.

In *Makers UK Limited v Mayor & Burgesses of the London Borough of Camden*[5] the solicitor for the referring party decided that the adjudicator appointed should be a solicitor because of legal issues arising in the case. He ascertained the identity of a particular solicitor who was on the RIBA panel. He did not seek to agree an adjudicator with the responding party. He contacted the solicitor adjudicator he had in mind to check his availability. Having spoken to that gentleman and ascertained his availability he applied to the RIBA to appoint him adjudicator. In doing so he advised the RIBA through the application form that "*Makers claim that Camden repudiated their Contract. That is a legal issue and is it suggested that Mr. Philip Harris of Wright Hassell be appointed if available*". A copy of that application form would have been furnished to the responding party. In a covering letter, which was not furnished to the responding party he also wrote "*Since the claim concerns the essentially legal issue of repudiation of contract, we respectfully invite you to nominate Mr. Philip Harris of Wright Hassell if available*".

Mr Harris was duly appointed as the adjudicator. When an application was made to enforce the adjudicator's decision, it was challenged on two grounds. Firstly it was pleaded that it was an implied term of the contract that neither party would seek to influence unilaterally the nominator's decision regarding the identity of an adjudicator by making unilateral representations to that body. The second ground was on the basis of apparent bias arising primarily from the unilateral contact made by the solicitor for the referring party with Mr Harris by phone in advance of his appointment. The court took the view that there was no obvious support in commercial or practical terms for implying such a term into the contract. At paragraph 29(5) of the judgment the court indicated that far from implying such a term, a contrary provision might be helpful:-

It is not necessarily wrong or unhelpful for a party to make representations. For instance, if the dispute is one which is very technical, say involving thermo-dynamics, it might be very sensible for the RIBA to be so informed. Similarly, if it was known that one or more of the people on the RIBA Panel of adjudicators were conflicted out, it would be sensible for the RIBA to be informed. It would be pointless for the RIBA to nominate someone who would either have to turn down the appointment or whose decision could be challenged on the grounds of bias.

The court found that there was no apparent bias arising out of the unilateral contact made by the solicitor for *Makers* with Mr Harris prior to his appointment. He held at paragraph 35(3):-

Whilst there is no positive encouragement for a party to contact a potential adjudicator to check his availability, there is no discouragement in the contract. Such contact if limited to checking availability or checking if there is any conflict is in itself unexceptionable and can be a sensible and practical step to take.

And at paragraph 35(5):-

There was nothing reprehensible in Dr Critchlow not mentioning to the RIBA or Camden that he had contacted for limited purposes the person whose name he was putting forward. Makers had no obligation to do so.

However the Court did offer guidance to the effect that it would be preferable that any such contact be limited and that where contact is necessary it should be in writing so that a full record of the communication is available (see page 154 below).

The recent judgment of Ramsey J in *Eurocom Limited v Siemens Plc*[6] underlines the necessity for a requirement that any communication by one party with the appointing body be furnished to the other party. In that case the party seeking the appointment advised the appointing body that a number of adjudicators were conflicted when they were not. The respondent did not become aware of this fact until after the adjudicator had furnished his decision. Ultimately the adjudicator's decision was quashed by reason of this deceit.

> 16   *The Chairperson and/or the Construction Contracts Adjudication Service of the Department of Jobs, Enterprise and Innovation may seek further information or clarification(s) from the applicant relevant to the nature of the dispute and such information or clarification(s) should be provided promptly by the applicant and copied to the other party/ parties to the payment dispute at the same time. No additional or other supporting information should be submitted by the applicant without a specific request for such information from the above-mentioned in this paragraph.*

The information sought by the Chairperson or the Adjudication Service is likely to be of a nature as would assist the Chairperson in identifying an appropriate adjudicator. This may give an applicant an opportunity to steer the Chairperson towards a particular individual or skill set. It is appropriate therefore that the respondent be furnished with any such communications. Whereas the applicant is prohibited from providing any further information without a specific request for it, the respondent is not so prohibited. A respondent may therefore comment upon any such additional information provided by the applicant.

> 17   *The Chairperson shall, following receipt of a completed application from a party to the construction contract made in accordance with paragraph 15 of this Code of Practice and subject to paragraph 16 of this Code of Practice, appoint an Adjudicator from the Panel.*

Presumably the appointment will only be made if the Chairperson is satisfied that the Notice of Intention was served in the manner required by the contract or the Act. This paragraph might suggest otherwise. It may be left to adjudicators to satisfy themselves as to proper service. If this were the case, it would hardly be necessary to provide details of service to the Chairperson under paragraph 15(iv).

> 18   *The appointment of an Adjudicator from the Panel shall be made by the Chairperson and notified in writing by the Construction Contracts Adjudication Service of the Department of Jobs, Enterprise and Innovation to the parties, normally within seven days after the receipt of the application to the Chairperson, subject to paragraph 16 of this Code of Practice. The date of the letter from the Construction Contracts Adjudication Service to the parties shall be deemed to be the date on which the appointment of the Adjudicator is made for the purposes of section 6(5) (a) of the Act.*

Earlier drafts of the Code of Practice did not qualify the requirement to appoint and notify, as is done here, by the insertion of the word "*normally*". The February

2016 draft did use that word but simply stated that the appointment would normally be made within seven days without reference to paragraph 16 or any similar qualification. It is not clear whether it is intended that the derogation from making the appointment within seven days is only to apply in the event of further information being sought under paragraph 16. It is thought more likely that the word "*normally*" is intended to apply generally. Presumably delay in the appointment where the application is made on Christmas Eve or during a holiday period would not invalidate the appointment. The extent of any delay or the circumstances of any delay in making and notifying the appointment would have to be considered in the context of what is normal, and what is acceptable. As indicated, the courts in the UK have applied time restraints contained in the legislation strictly on the basis that Parliament obviously intended very strict time limits to apply and this intention should be respected. This may not be the case in Ireland, at least in so far as the appointment by the chairperson is concerned. Delay in making the appointment by the chairperson may arguably never invalidate the appointment.

Certainty as to the date on which the appointment is made is necessary given the requirement to serve the referral within a further seven days. If the precedents set by the courts in the UK apply, that time limit will be strictly applied.

It is questionable whether sub-section 10(3) applies to the notification given by the chairperson under the Code of Practice. That sub-section provides that where notice is served under the Act the time for service will be extended where the last day falls on a Saturday, Sunday or public holiday.

> 19   *An Adjudicator appointed by the Chairperson to a payment dispute shall, within two days of such appointment, request of the parties in writing to disclose any information indicating any potential conflict of interest that may arise from the person's appointment as Adjudicator. He/she shall draw the attention of the parties to the provisions of paragraph 32 of this Code of Practice. The Adjudicator shall at the same time provide the parties with his/her terms and conditions of appointment, including the basis for his/her fees, costs and expenses.*

This provision is almost identical to paragraph 9 which is discussed above. The same points arise here as in relation to it.

> 20   *If the information disclosed indicates a potential conflict of interest, the Adjudicator may only proceed with the adjudication where he/she is satisfied that the disclosures are frivolous or vexatious; that no professional or ethical concerns arise; and that no actual conflict of interest exists.*

Given the relatively small size of the construction industry in Ireland, many adjudicators on the Minister's panel represent or advise parties in relation to disputes when not acting as adjudicators, it would usually not be difficult to find a potential conflict of interest in almost any case. This paragraph replaces paragraphs 20 and

21 of the Statutory Instrument of the 5th July 2016. This was a necessary change because that version of the Code of Practice required the parties to consent to the appointment proceeding once a potential conflict of interest was identified. This in effect gave a reluctant respondent the right to veto an appointment or even a series of appointments. This new paragraph 20 requires the ultimate decision to be made by the appointee. The earlier paragraph 20 contained a sentence which has now been omitted: "The adjudicator shall also advise the Chairperson of such potential conflict of interest". It might have been appropriate to retain that provision as one can see a worthwhile role for the chairperson in offering objective advice to the appointee as to whether a potential conflict might be regarded as an actual conflict.

Referral of a Payment Dispute to an Adjudicator

21 *Following the appointment of an Adjudicator, the Referring Party shall in accordance with section 6(5) of the Act refer the payment dispute to the Adjudicator in writing within seven days of the Adjudicator's appointment, and the Referring Party shall provide a copy of all such documentation to the Responding Party/Parties at the same time.*

22 *The referral of the payment dispute to the Adjudicator shall include:*

   (i)   *the name, address and contact details of each party to the construction contract;*
   (ii) *relevant details of the payment dispute to include the amount in dispute (even if the amount is zero), the nature of the payment dispute, and the site address;*
   (iii) *a copy of the Notice of Intention including any accompanying documents attached to that Notice;*
   (iv) *the date when the Notice of Intention was served on the Responding Party/Parties and how this was done;*
   (v)   *the contentions on which the Referring Party intends to rely upon to support their case; and*
   (vi) *relevant details to identify the construction contract and any supporting information that may assist an Adjudicator in understanding the nature of the payment dispute. Where a written construction contract exists, this must be attached.*

Paragraph 21 does no more than repeat what is already in section 6(5) of the Act.

   Items (i), (ii) and (vi) are exactly the same as items (i), (ii) and (iv) in the list of items to be included in a Notice of Intention under paragraph 5. Therefore these provisions are superfluous in that they are included in item (iii).

   Item (iv) would be particularly important where the appointment was made by the agreement of the parties. Where the appointment is made by the Chairperson, presumably the Chairperson or the Adjudication Service will have checked that service was completed in accordance with the contract and/or the Act.

The new and important element contained in the referral is item (v) whereby referring parties must set out their contentions in support of the case being made. Presumably the referring party should also include at this stage any additional documents upon which it relies. It is perhaps surprising that the list does not provide for such additional documents but it could hardly be successfully argued that there is no entitlement to introduce additional documents at this stage. It may well be argued that the referring party should not be entitled to introduce further documentation at a later stage that it could or should have included in the referral. Paragraph 7(2) of the Scheme operable in the UK states that a referral notice shall be accompanied by such documents as the referring party intends to rely upon. Although unstated, that presumably is required in Ireland also.

A referral will not be valid unless the contentions required by this paragraph are included. It is unlikely that the courts will be exacting in their requirements for compliance with 22(v) but a total lack of compliance with that requirement would presumably invalidate the referral. It is to be noted that paragraph 22(v) goes beyond what is required in the UK. The schedule to the Scheme at paragraph 1 sets out the required content of the notice of adjudication and at paragraph 7(2) it sets out the requirements of the referral notice. Neither provision expressly requires the claimant / referring party to set out the contentions on which it relies. Given the purpose of this legislation and the assumed intention that the adjudication process will be availed of without the assistance of professional advisers, it is likely that a very basic statement of the contentions relied upon will suffice and that the referring party will not be rigidly held to these contentions if it later offers different or other arguments. However these are uncharted waters and it remains to be seen how the courts will treat this requirement.

In relation to timescales the TCC has indicated that it is not prepared to impeach the jurisdiction of an adjudicator for minor overruns of time in the delivery of the documents relied upon, as opposed to the referral itself. In *PT Building Services Limited v ROK Build Limited*[7] Ramsay J addressed the issue of documents that should have been included with the referral being served outside the time required for delivery[8]. The referral itself was delivered on time but some of the associated documents were not delivered until the next day. Those documents were not vital in terms of the adjudicator being able to consider the matter upon receipt of the referral and in those circumstances the Court found that the referral was valid.

Adjudication of a Payment Dispute – Procedures and Decision

23 *The Adjudicator in any payment dispute under the Act shall be impartial, independent and only adjudicate where satisfied that no actual conflict of interest exists.. He/she shall observe the principles of procedural fairness, which shall include giving each party a reasonable opportunity to put their case and to respond to the other party's case.*

The requirement to act impartially requires the adjudicator to avoid contact with one party alone. The adjudicator is very much in the same position as a judge or arbitrator in this regard. If one party is given an opportunity to discuss the dispute with the adjudicator without the other party being present, this may give rise to a presumption that the principles of natural justice have not been adhered to. It is rarely that one can point to a specific provision in the decision, judgment or award to establish that anything occurred at any such meeting which proved to be material to the determination of the dispute. However the very fact that the decision maker was unilaterally exposed to the views of one party may well give rise to an assumption of such a breach. Natural justice requires that parties be entitled to know the case they have to defend and be given an opportunity to defend it. If they do not know what is being said to the adjudicator, they cannot defend it. In *Discain Project Services Limited v Opecprime Developments Limited*[9] the adjudicator had two private conversations with the employees of one of the parties. The adjudicator did not contact the employees but rather was contacted by them. Nonetheless he proceeded to have conversations with them which were not communicated to the respondent. The Court refused to enforce the adjudicator's determination in the circumstances notwithstanding that there was no evidence that the adjudicator was influenced by what these employees had said.

In *Glencot Development & Design Company Limited v Ben Barrett & Son (Contractors) Limited*[10] Judge Humphrey Lloyd of the Technology & Construction Court took the opportunity to review the current law on bias in a comprehensive manner. In that case the parties had almost reached an agreement immediately prior to the adjudication commencing. The adjudicator was asked to mediate on the outstanding issue, it being understood that if negotiations broke down, he would resume his role as adjudicator. Discussions then took place over the next six hours. In his role as mediator the adjudicator met with the parties separately and together. When the negotiations broke down the adjudicator wrote to the parties indicating his intention to resume his role as adjudicator but requesting that if either party considered that his position had been compromised by being involved as a mediator he should be informed immediately with a view to his withdrawing from the adjudication. The respondent failed to respond immediately to this but did, on resumption of the adjudication, belatedly indicate that in its view the adjudicator was compromised and should not proceed. The adjudicator took legal advice and decided to proceed with the adjudication. His resulting decision was challenged in the TCC Court. That challenge was upheld. The essential finding of the Court is to be found in paragraph 24:-

> In this case Mr Kennedy submitted that there was no evidence that anything emerged in the discussions that might have affected Mr Talbot's decision or approach. That very submission effectively makes the defendant's case. Whilst in an adjudication it is permissible to make inquiries and receive evidence and submissions from one party alone there is a clear obligation on the adjudicator to give any absent party a complete and accurate account of what has taken place. Mr Talbot went to and fro between the parties. We do not know what he heard or learned.

It is sometimes impossible for adjudicators to avoid a party making contact directly with them. Obviously adjudicators should do whatever they can to avoid such contact but if it occurs, it is essential that adjudicators communicate to the other side exactly what was said to them in the unwarranted communication. Provided they do this, they cannot be criticised. If the adjudicator fails to do so, and material aspects of the case, as opposed to an issue such as the date for the hearing, are discussed, the failure to communicate in that manner may give rise to a breach of natural justice. It is submitted that even if the contact made by one party is simply to ascertain details such as the venue, date or time for a meeting, the fact that it occurred should be recorded by the adjudicator by way of a note to the other side simply recording the nature of the query and the fact that no further conversation took place. If this is done at the time, there can be no misunderstanding. If it is not done at the time and the other party subsequently learns that there was a direct conversation between the adjudicator and the opposition of which he or she was not aware, doubts about the impartiality of the adjudicator may fester notwithstanding the apparent insignificance of the subject matter discussed.

The concepts of impartiality, bias and independence are theoretically distinct but in practice are interlinked and overlap. It is to be noted that the Scheme requires only that the adjudicator acts impartially[11]. The Scheme makes no mention of bias or independence. Generally however, all three concepts are considered in the case law of the courts and by the authorities under the generic term of bias. Bias can arise in many different forms. It may arise as a result of some wrongdoing on the part of the adjudicator, or on the part of a party, but does not necessarily involve wrongdoing or even moral turpitude on the part of anyone. Furthermore the issue may arise before or after the adjudicator has made a decision.

The question is rarely one of whether the adjudicator was actually biased – it is usually a question of whether on a balanced view, a reasonable person might suspect that the facts might give rise to bias. This test has been described in different ways. In *Ellis Building Contractors Limited v Vincent Goldstein*[12] Mr Justice Akenhead suggested that one must ask oneself *"Whether, on an objective appraisal, the material facts give rise to a legitimate fear that the adjudicator might not have been impartial"*.

Such an issue may arise where the adjudicator has expressed strong views, perhaps through academic articles, on issues relevant to the case prior to it being referred to him for adjudication. For instance, the adjudicator may have published papers on how claims for extensions of time should be proven. It is submitted that generally academic views of this nature would not be sufficient to give rise to any suspicion of bias[13]. However if the adjudicator's views were quite extreme and were expressed so robustly as to convey the impression that he could not approach the adjudication with an open mind, a presumption of bias might well arise.

The case of *Dublin Well Woman Centre Limited v Ireland*[14] was listed for hearing in the High Court before Ms. Justice Carroll. The issue at the heart of the case was whether the Plaintiff should be entitled to distribute information regarding abortion services available in other countries. The Judge was chairwoman of the Commission for the Status of Women and in that capacity had made, in conjunction with others, a submission to the Government on the issue of the availability of information regarding abortion. At the outset the Judge was asked to recuse herself by reason of the perceived bias arising from these facts. She refused to do so but that decision was overruled by the Supreme Court, that Court pointing out that "It is of particular importance that neither party should have any reasonable reason to apprehend bias in the Courts of Justice".

It is probably the case that the courts would be more exacting in the standard applied to apparent bias where the issue comes before the court prior to the hearing than would be the case if the outcome of a hearing was challenged on the basis of bias. Ms. Justice Denham in the *Dublin Well Woman* case went on to say "Further, once the question of a possible perception of bias has been raised reasonably on the grounds of pre-existing non-judicial position and actions, it would be contrary to constitutional justice to proceed with a trial". Had no application been made to the Judge to recuse herself and the case had proceeded but the outcome was challenged on the basis of bias on the part of the Judge, it does not follow from the decision made in the actual case that the challenge would have been successful.

The Court of Appeal in *Shannon v Country Casuals Holdings Plc*[15] upheld the decision of the High Court Judge to continue a trial notwithstanding, just before the closing speeches were to be made, he was shown what had been proposed in terms of settlement. He decided to do so on the basis that he considered that he would not be influenced by this disclosure. He therefore applied the test of actual bias as opposed to apparent bias. It is telling however that in doing so he said "I have no hesitation in expressing the view that I can conclude this trial without any prejudice, either conscious or indirect. I have heard the evidence. It would be quite different if we were in the early stages of the trial but we are not". Had therefore the Judge seen the terms of settlement before the trial commenced, or at an early stage, he may well have recused himself on the basis of apparent bias. Having regard to the fact that the case was almost concluded, he applied the subjective test of actual bias.

In Ireland, the test for objective bias was more recently restated in the case of *Craig v An Bord Pleanala*[16]. In that case Hedigan J stated:

> In its place was raised a claim of objective bias. This involves the reasonable apprehension of a reasonable person fully informed of all the relevant facts that he would not have a fair hearing from an impartial tribunal. Denham J in Bula Ltd v Tara Mining Ltd (No 6)[17] approached the issue as follows:
>
> . . . it is well established that the test to be applied is objective, it is whether a reasonable person in the circumstances would have a reasonable apprehension that the Applicants would not have a fair hearing from an impartial judge on the issues. The test does not invoke the apprehension of the judge or judges. Nor does it invoke the apprehension of any party. It is an objective test - it invokes the apprehension of the reasonable person.

Fennelly J in *Kenny v TCD*[18] stated:

20 The hypothetical reasonable person is an independent observer, who is not over sensitive, and who has knowledge of the facts. He would know both those which tended in favour and against the possible apprehension of a risk of bias . . . .

21 The test of objective bias is expressed in general terms. Its application demands an appreciation of all the circumstances of the individual case, followed by a particularly careful exercise of the faculty of judgment. In his judgment in O'Neill v Beaumont Hospital Board [1990] ILRM 419, where the allegation was one of pre-judgment bias, Finlay CJ expressed the view, at p 439, that, in analysing the facts, he should 'take the interpretation more favourable where there is ambiguity to the Plaintiff than to the Defendant.' Whether or not that is a principle of general application, it applies in a special way in the present case, where this court is asked, in a very real way, to adjudicate on whether one of its own judgments was tainted by objective bias. That fact obliges it, in order to ensure respect for the principle that justice must not only be done but be seen to be done, to act with great care and circumspection. It should err on the side of caution.

Some of the issues arising from bias that have been tested by the courts in the context of adjudication have arisen from without prejudice documentation being put before the adjudicator. In *Volker Stevin Limited v Holystone Contracts Limited*[19] Mr Justice Coulson found that the fact that the adjudicator became aware of the fact that a without prejudice offer had been made, but not of the amount of the offer, did not affect his decision and accordingly there was no bias. In the case of *Ellis Building Contractors Limited v Vincent Goldstein*[20] the respondent had made a without prejudice offer of settlement prior to the adjudication. The claimant not only referred to the fact that an offer had been made but furnished the letter of offer, with the sum offered redacted, to the adjudicator. Surprisingly, Mr Goldstein did not object to this at the time. However it did challenge the adjudicator's decision on the grounds of bias in the Technology & Construction Court. The without prejudice offer was of course privileged in its entirety and should never have been mentioned to the adjudicator, let alone provided to him. It was not just the amount of the offer that was privileged – as is always the case, it was the very fact that an offer was made that was privileged. The Judge found however that there was no apparent bias on the part of the adjudicator arising from the fact that he was furnished with the letter and upheld the adjudicator's decision. Had the adjudicator's decision given any hint that the fact of the offer had influenced any material aspect of the decision, it is submitted that the outcome would have been different.

Cases rarely come before the Court on an issue of bias before a decision is made by the party against whom bias is alleged. The *Dublin Well Woman* case was an exception to this. Usually the Court has to deal with the issue after the decision has been made. The issue for the Court is invariably a consideration of whether there has been actual bias, and, even if there has not been, and the same is rarely alleged, whether there is a reasonable apprehension of bias. The House of Lords in *Regina v Gough*[21] seemed to favour the test of actual bias.

More recent judgments, reinforced by the Court of Human Rights in *Gregory v United Kingdom*[22] have tended to give prominence to the objective test of whether an observer, informed of the relevant facts, might reasonably apprehend that the tribunal would be prejudiced. The applicable principles as laid down in *Gregory* were summarised by Humphrey Lloyd J in *Glencot as follows:-*

> 83   *In Gregory v United Kingdom (1997) 25 E.H.R.R. 577 the Court recognised that it was possible for risk of prejudice on the part of a jury to be effectively neutralised by an appropriate direction from the judge. The Court commented at p.595 that the legal principles applied in England corresponded closely to its own case law on the objective requirements of impartiality.*
>
> 84   *We would summarise the principles to be derived from this line of cases as follows:*
>
> (1)   *If a Judge is shown to have been influenced by actual bias, his decision must be set aside.*
>
> (2)   *Where actual bias has not been established the personal impartiality of the Judge is to be presumed.*
>
> (3)   *The Court then has to decide whether, on an objective appraisal, the material facts give rise to a legitimate fear that the Judge might not have been impartial. If they do the decision of the Judge must be set aside.*
>
> (4)   *The material facts are not limited to those which were apparent to the applicant. They are those which are ascertained upon investigation by the Court.*
>
> (5)   *An important consideration in making an objective appraisal of the facts is the desirability that the public should remain confident in the administration of justice.*

In *Makers UK Limited v Mayor & Burgesses of the London Borough of Camden*[23] Mr Justice Akenhead had to consider whether there was apparent bias on the part of an adjudicator where he had been contacted prior to the adjudication by one of the parties solicitors, had confirmed his availability and that solicitor had subsequently written to the institution suggesting, without furnishing a copy to the other side, that the particular adjudicator to whom he spoke would, because of his legal skills, be appropriate for the adjudication, and that adjudicator was duly appointed. The judge concluded that there was no bias but saw fit to conclude his judgment by making the following observations for the guidance of parties involved in adjudication:-

(1)   *It is better for all concerned if parties limit their unilateral contacts with adjudicators both before, during and after an adjudication; the same goes for adjudicators having unilateral contact with individual parties. It can be misconstrued by the losing party, even if entirely innocent.*

(2) *If any such contact, it is felt, has to be made, it is better if done in writing so that there is a full record of the communication.*

(3) *Nominating institutions might sensibly consider their rules as to nominations and as to whether they do or do not welcome or accept suggestions from one or more parties as to the attributes or even identities of the person to be nominated by the institutions. If it is to be permitted in any given circumstances, the institutions might wish to consider whether notice of the suggestions must be given to the other party.*

The question of bias in relation to adjudication can also arise where adjudicators issue their preliminary views and where it could be alleged that the adjudicators have made up their minds prior to issuing their final decisions. This issue was considered in the case of *Lanes Group Plc v Galliford Try Infrastructure Limited*[24]. In siding with the adjudicator and following the 'fair-minded observer' test as set out *In re: Medicaments and Related Clauses of Goods (No 2)*[25] Jackson LJ considered that the expression of preliminary views created an opportunity to correct errors and concentrate minds.

The obligation to observe the principles of procedural fairness, including giving each of the parties a reasonable opportunity to put their case, is dealt with in Chapter 13 in the context of constitutional issues.

> 24   *For the purposes of the adjudication proceedings, the Adjudicator may:*
>
> (i)   *request any reasonable supporting or supplementing documents pertaining to the payment dispute detailed in the Notice of Intention and/or in the referral of the payment dispute to the Adjudicator;*
>
> (ii)   *take the initiative in ascertaining the facts and matters required for a decision and make use of his/her own specialist knowledge, if it is appropriate to do so. If the Adjudicator uses any such specialist knowledge he/ she shall disclose this to the parties as appropriate;*
>
> (iii)   *appoint experts, assessors or legal advisers, provided that the parties have been notified of their identity and their terms of reference;*
>
> (iv)   *make site visits and inspections or carry out tests, subject to prior notification to the parties and obtaining any necessary consent from a third party or parties;*
>
> (v)   *invite written submissions/representations and evidence from the parties, if appropriate;*
>
> (vi)   *meet jointly with the parties and their representatives, if any, to enable further investigation;*
>
> (vii)   *hold a teleconference with the parties, with the consent of the parties; and*
>
> (viii)   *hold an oral hearing, where appropriate.*

An earlier draft of the Code of Practice set out the powers of the adjudicator in almost identical terms to those set out at paragraph 13 of the Scheme. Paragraph

24 is quite different to paragraph 13 of the Scheme. A number of powers have been omitted and new ones have been added. It is doubtful however that there is any great significance to this. For instance, paragraph 13 of the Scheme entitles the adjudicator to "*meet and question any of the parties to the contract and their representatives*". This was included in an earlier draft of the Code of Practice. Its ultimate omission hardly indicates that the adjudicator is not entitled to question the parties and their representatives. Under Section 6(9) of the Act the adjudicator may take the initiative in ascertaining the facts and any of the powers set out in paragraph 13 of the Scheme, but omitted from paragraph 24 of the code, it is submitted, are inferred by the primary legislation.

Neither the Act nor the Code of Practice gives any guidance as to the level of proof an adjudicator may require of the parties. Obviously adjudicators would be entitled to rely upon copies of documents without proof of those documents save in exceptional circumstances where the validity of an important document is challenged. Whereas in court proceedings evidence would undoubtedly be required to prove such a document it is questionable whether even in those circumstances an adjudicator would be required to hear oral evidence. If there was such a requirement, this might provide a respondent who is seeking to derail an adjudication the opportunity to do so. It may be that even where an important document is challenged, an adjudicator, whilst taking account of the challenge, may nonetheless reach the conclusion that the document is likely to be valid without hearing evidence. In *Intero Hospitality Projects Pty Limited v Empire Interior (Aust) Pty Limited*[26], Muir JA observed at paragraph 51:-

> *It is apparent also that in making determinations under the Act adjudicators will often lack the evidence upon which and the time within which to make fully informed considered determinations. That does not matter in the scheme of things as adjudicators' determinations do not finally determine the parties' contractual rights. That is left to the Courts or to the alternative dispute resolution process agreed upon by the parties.*

Paragraph 24(i) must be read in conjunction with paragraph 32 of the code which entitles the adjudicator to draw adverse inferences if a party fails or refuses to provide documents requested by the adjudicator.

The entitlement of the adjudicator to make use of his or her own specialist knowledge must be exercised with caution. Clearly adjudicators do avail of their expertise in gaining an understanding of the issues in dispute and that is both desirable and permissible, but the adjudicator must not avail of specialist knowledge in making a decision where the parties have not been given the opportunity to test or comment upon the accuracy or relevance of that knowledge (see page 158).

As indicated, it is anticipated that adjudicators will rarely avail of the entitlement to appoint experts, assessors or legal advisers. Any advice given by such individuals to the adjudicator would have to be shared with the parties and the parties would have to be given an opportunity to respond to it. The time restraints simply do not permit of such luxuries. This fact adds weight to the requirement set out in paragraph 6(ii) of the Code of Practice that adjudicators consider at the outset whether disputes are within their competence.

Whether or not oral hearings prove more prevalent in Ireland than elsewhere remains to be seen. Whilst adjudicators do have the power to take evidence and conduct oral hearings most are conducted without a hearing. The 12th Report of the Adjudication Reporting Centre covering the year to April 2012[27] concluded in this respect:-

> The majority of adjudications were conducted on a 'documents only' basis. This might be due to speed and convenience, avoidance of claims of procedural error or bias or it could be due to the preference of the adjudicators to avoid hearings which are traditionally within the comfort zone of the lawyers. The use of an interview procedure with both parties present almost halved since the last report, while the use of full hearings rose from 3.1% to 11.5% of the total.

> 25 *The Adjudicator shall upon receipt of the referral of a payment dispute from the Referring Party, inform the parties in writing of the date on which it was received by the Adjudicator. This date of receipt shall be regarded as the date on which the referral of the payment dispute to the Adjudicator has been made for the purposes of section 6(6) of the Act.*

This gives certainty to the date from which the time for the making of the adjudicator's decision is to run. The Act provides that the twenty-eight day period begins with the date on which the referral is made.

> 26 *The Adjudicator shall, at the same time, also inform the parties of the procedures that he/she intends to apply during the adjudication process. This shall include directions as to the timetable for the adjudication and any deadlines to be adhered to by the parties and/or limits as to the length of written documents. The Adjudicator shall draw the attention of the parties to the provisions of paragraph 32 of this Code of Practice. The Adjudicator may revise his/her guidance to the parties on the above mentioned matters in circumstances where he/she considers it necessary to do so and he/she shall inform the parties of any such change as appropriate.*

The first sentence confirms the entitlement of the adjudicator to dictate the required procedures. No doubt the adjudicator will respect the wishes of the parties but the adjudicator is not bound by those wishes. This provision must be read in conjunction with paragraph 27 which obliges the adjudicator to ensure that the procedure adopted is commensurate with the nature and value of the payment dispute and also to be mindful of whether or not an oral hearing is required.

The fact that the adjudicator must set out the procedures and timetable at the same time as confirming the date upon which the referral was received diminishes the opportunity to the adjudicator to provide tailor made procedures for the specific dispute. This requirement is likely to result in adjudicators operating to routine procedures and deadlines. In the UK an expectation has arisen through experience that normally adjudicators will allow fourteen days for the respondent to reply to the referral, and a further seven days for the referring party to reply to that submission, thus leaving the adjudicator with seven days to prepare the decision. In practice such directions rarely apply without qualification as issues invariably arise in respect of which parties feel obliged or entitled to comment. The reference here to paragraph 32 of the Code of Practice, which deals with a party failing to comply with the adjudicators directions, may give this provision added weight.

Irrespective of such limitations as the adjudicator may seek to impose in terms of deadlines and lengths of submissions, the adjudicator must meet the requirements of natural justice and is therefore always obliged to provide an opportunity for the other party to respond to new material no matter how late in the proceedings that material is received. If new material is received from one party so late that the other party cannot respond to it, it may be that the new material should be disregarded. However if that material is to be disregarded, it may mean that earlier submissions to which it relates also have to be disregarded to avoid the adjudicator making a decision on a chain of evidence which is clearly incomplete. A judgment call must be made by the adjudicator depending upon the circumstances of each individual case and the significance of the new material.

The adjudicator's timetable and directions should reflect an intention on the part of the adjudicator to avoid drift or time lapses. The adjudicator should be concerned to keep tight control over the timetable throughout the process. If the adjudicator feels at this stage that it will be essential that extra time be given for the process it is at this stage that the adjudicator should raise the matter. Hesitancy and indecision on the part of the adjudicator are to be avoided.

> 27   *The Adjudicator shall ensure that the procedure adopted is commensurate with the nature and value of the payment dispute and he/she shall be mindful of whether or not an oral hearing is required having regard to matters such as to whether or not there is a conflict of fact or other relevant matter that requires such a hearing.*

This provision must be read in conjunction with paragraph 23 which requires the adjudicator to observe the principles of procedural fairness. The encouragement here to adopt shortcuts where appropriate having regard to the monetary value of the dispute and the entitlement under paragraph 26 to limit the length of written submissions and to set down deadlines cannot be availed of to deprive a party of natural justice.

In the case of *CJP Builders Limited v William Verry Limited*[28] the adjudicator decided that the adjudication rules were such that he refused to consider the claimant's response which had been served a few hours late under the applicable rules. In his judgment, Akenhead J stated:

83   Reliance was placed upon my observations in Cantillon Ltd v Urvasco Ltd [2008] BLR 250 at paragraph 57:

> 'From this and other cases, I conclude as follows in relation to breaches of natural justice in adjudication cases:
>
> (a) it must first be established that the adjudicator failed to apply the rules of natural justice;
> (b) any breach of the rules must be more than peripheral; they must be material breaches;
> (c) breaches of the rules will be material in cases where the adjudicator has failed to bring to the attention of the parties a point or issue which they ought to be given the opportunity to comment upon if it is one which is either decisive or of considerable potential importance to the outcome of the resolution of the dispute and is not peripheral or irrelevant;
> (d) whether the issue is decisive or of considerable potential importance or is peripheral or irrelevant obviously involves a question of degree which must be assessed by any judge in the case such as this;
> (e) it is only if the adjudicator goes off on a frolic of his own, that is wishing to decide a case upon a factual or legal basis which has not been argued or put forward by either side, without giving the parties an opportunity to comment or, where relevant put in further evidence, that the type of breach of the rules of natural justice with which the case of Balfour Beatty Construction Co Ltd v The Camden Borough of Lambeth was concerned comes into play. It follows that, if either party has argued a particular point and the other party does not come back on the point, there is no breach of the rules of natural justice in relation thereto.'

The Cantillon case was concerned with an allegation that the adjudicator had made his decision upon a factual or legal basis not argued or put forward by either side without giving the parties opportunity to comment or address. That is not the case here . . . .

85   As to whether the breach was a material one, on analysis the only point put forward by CJP is that the Adjudicator's decision in the second adjudication demonstrates that the Adjudicator would probably have found against Verry even if he had considered the Response in the first adjudication. CJP argues that the onus of proof must be on Verry to show that the breach was a material one and that the adjudicator would have reached a different decision. I am satisfied that the breach here was a material breach:

(a) In itself, the failure to disregard the whole of Verry's response both as to argument and as to evidence was and must have been material. There comes a point when a breach of the rules of natural justice is so pervasive that the only proper conclusion to come to is that the breach is material.

(b) It is not necessary for the court to go so far as having to investigate the facts to determine whether the adjudicator would have reached a different decision in substance if he had considered the Response. All one need say (and I do) is that there was a real possibility that the adjudicator could have reached a different decision. I am satisfied that there is a real (as opposed to fanciful) possibility in this case.

(c) Because Verry decided to stop participating in the second adjudication part way through, there is no telling what the adjudicator would have decided if Verry had pressed its case with force and logic throughout the remainder of the adjudication.

It is thought that in Ireland it is likely that the courts will follow the rationale of Akenhead J as set out above and will therefore uphold the applicability of principles of natural justice where these are in conflict with procedures laid down by the relevant contract between the parties or by the adjudicator.

The obligation on the adjudicator to be mindful of whether an oral hearing is required, and the wording of that provision is to be contrasted with paragraph 19 of the last draft which stated:-

The Adjudicator shall use reasonable endeavours to process the payment dispute between the parties in the shortest time and at the lowest cost and shall, in order to achieve this, only hold an oral hearing where there is a genuine conflict of fact or other situation requiring such a hearing.

It could be said that paragraph 19 of the draft discouraged adjudicators from availing of oral hearings whereas paragraph 27 of the code encourages them to do so. This may be reflective of rather jaundiced comments of various judges and former judges of the High and Supreme Courts on the operation of adjudication in Ireland having regard to constitutional rights. The extent to which an oral hearing should be permitted will undoubtedly be a key feature in future court decisions on constitutional issues relating to this legislation and the matter is dealt with in that context in Chapter 13.

28 *The Adjudicator shall use reasonable endeavours to process the payment dispute between the parties in the shortest time and at the lowest cost. He/she shall promptly notify the parties of any matter that will slow down or increase the cost of making a determination.*

The obligation to complete the process in the shortest time is significant. This obligation should discourage adjudicators from seeking additional time merely to suit the adjudicator's own convenience. Adjudicators should seek additional time to provide a decision only if the nature of the dispute so requires. The related obligation to complete the process at the lowest cost begs the question as to whether a meeting with the adjudicator, let alone hearings should be conducted (in the UK, the word *"hearing"* is rarely used in the context of adjudication. The word *"meeting"* is used even where the purpose of the meeting is to hear oral evidence). Traditionally arbitrators, conciliators and mediators tend to arrange preliminary meetings to discuss how the process is to be conducted. Such meetings are considered in the UK as being unnecessary and rarely occur. Most adjudications are conducted without any meeting of the parties with the adjudicator at any time. If meetings are held, it is usually for the purpose of taking oral submissions on specific points or hearing very limited evidence. Although adjudicators have the power to engage experts to assist them, this is rarely done simply because of the time restraints. Having regard to paragraph 28 an adjudicator should also take account of the cost of any initiative of that nature. The obligations imposed by paragraph 28 in terms of time and cost will no doubt be welcomed by adjudicators as they may justify an approach that would, in any event, be preferable to efficient adjudicators.

The obligation to notify the parties of any matter that would slow down or increase the cost of making a determination is laudable. An adjudicator who has been bombarded with documents and submissions may feel that this is unhelpful to the process and is only adding to the cost. Under this provision not only would adjudicators have an entitlement to raise such a point with the parties, they may have an obligation to do so.

29 *The parties may agree to revoke the appointment of the Adjudicator in accordance with section 6(18) of the Act and shall be jointly and severally liable for the payment of the reasonable fees, costs and expenses incurred by the Adjudicator up to the date of the revocation.*

This paragraph is unnecessary. It merely repeats what is already in the Act.

30 *In accordance with section 6(17) of the Act, the Adjudicator may resign for reasonable cause at any time on giving notice in writing to the parties to the payment dispute. Such resignation, anonymised in terms of the details of the parties to the dispute, shall be notified by the Adjudicator to the Construction Contracts Adjudication Service of the Department of Jobs, Enterprise and Innovation for the purpose of compiling statistical information relating to the Act.*

Although the information to be provided is anonymised in respect of the parties to the dispute, the identity of the adjudicator resigning will be known to the Adjudication Service. No doubt the Chairperson will require an explanation from an adjudicator who is a member of the panel.

The first sentence gives the impression that Section 6(17) of the Act only allows an adjudicator to resign *"for reasonable cause"*. This is not the case. Section 6(17) says an adjudicator may resign at any time on giving notice in writing to the parties. The adjudicator is not obliged to give reasons or to establish reasonable cause under the Act.

31   *Upon such resignation the adjudication is at an end subject to the pay-ment by the parties of the reasonable fees, costs and expenses incurred by the Adjudicator up to the date of resignation. The parties shall be jointly and severally liable for the payment of the reasonable fees, costs and expenses incurred by the Adjudicator up to the date of resignation.*

The first sentence states no more than the obvious and the second merely repeats what is already in the Act.

32   *If a party to the adjudication, without showing sufficient cause, fails to:-*

(i)   *attend a meeting; or*

(ii)   *comply with any directions of the Adjudicator made in accordance with paragraph 28 of this Code of Practice; or*

(iii)   *disclose any information indicating a potential conflict of interest as required to do in accordance with paragraphs 9 and 19 of this Code of Practice; or*

(iv)   *produce any document or written statement requested by the Adjudicator; the Adjudicator may:-*

a)   *continue the adjudication in the absence of a party;*

b)   *continue the adjudication without the document or written statement requested;*

c)   *draw such inferences from that failure to comply as circum-stances may, in the Adjudicator's opinion, be justified;*

d)   *make a decision on the basis of the material properly provided; and*

e)   *make a decision apportioning the fees, costs and expenses of the Adjudicator, as appropriate.*

This was obviously considered to be a very important provision by the draftsman of the code. The adjudicator is obliged under paragraphs 9 and 19 to draw the attention of the parties to this provisions at the very outset of the adjudication and is again obliged to draw the attention of the parties to it when issuing directions under paragraph 26. The powers conferred by this paragraph may not warrant such attention. It is thought that the adjudicator would have all of these powers by the very nature of the process even if they were not specifically set out in the leg-islation. The very fact that the process must be conducted in such a short period of time, means by necessity that the adjudicator must proceed ex parte if necessary. It is thought that the entitlement of the adjudicator to take into account the non-cooperation of a party in the allocation of the adjudicator's fees and expenses does not have to be stated any more than the unstated entitlement of the adjudicator to take into account time wasted by a party.

The powers of the adjudicator as set out above are almost identical, word for word, to paragraph 15 of the Scheme with two exceptions. The equivalent to item (d) in the Scheme states "*Make a decision on the basis of the information before him attaching such weight as he thinks fit to any evidence submitted to him outside any period he may have requested or directed*". The other difference is that item (e) is omitted from paragraph 15 of the Scheme. The reason for the first of these differences may arise from paragraph 17 of the Scheme which requires adjudicators to consider any relevant information submitted to them without limitation. The inclusion of the words "*properly provided*" in item (d) of paragraph 32 may infer that if the material was provided outside of the adjudicator's directions it need not be considered. Given that paragraph 32 is almost an exact copy of paragraph 15, and is clearly based upon it, one is almost driven to this conclusion. However the Code of Practice cannot affect the requirements of natural or constitutional justice and it would be rarely that an adjudicator could justify ignoring or refusing to read a document provided outside of the directed submissions.

> 33   *The Adjudicator shall, in accordance with the Act, reach a decision within 28 days beginning with the day on which the referral is made or such longer period as is agreed by the parties after the payment dispute has been referred. The Adjudicator may extend the period of 28 days by up to 14 days, with the consent of the Referring Party.*

This adds nothing to what is already stated in the legislation at sub-sections 6(6) and 6(7).

> 34   *The decision of the Adjudicator on a payment dispute shall be in writing and it shall be signed and dated by the Adjudicator. Unless the parties agree otherwise in writing, the decision shall include the reasons for the decision.*

The Scheme at paragraph 22 only requires the adjudicator to provide reasons if requested by one of the parties to do so. In practice adjudicators decisions invariably include reasons and it is most unlikely that the parties would agree otherwise. The reasons given should demonstrate that the adjudicator understood the issues and the arguments made by the parties in respect of them. They should also demonstrate why the adjudicator preferred one argument over another, at least in so far as major issues are concerned.

If adjudicators fail to give reasons when they are required to do so, obviously their decisions will be unenforceable. Where, however, the adjudicator purports to give reasons but the reasons given are unintelligible or otherwise defective the enforceability of the decision will depend upon the nature and extent of the defect.

It is clear from the authorities that the standard of reasoning required of an adjudicator is not that required of a judge or even an arbitrator. The courts recognise that an adjudicator is obliged to often make complex decisions within a very short period of time and, as a result, the expression of those decisions may not be as clear as would otherwise be expected.

The following judicial pronouncements in the UK identify the approach taken by the courts:-

> *If an adjudicator is requested to give reasons pursuant to paragraph 22 of the Scheme, in my view a brief statement of those reasons will suffice. The reasons should be sufficient to show that the adjudicator has dealt with the issues remitted to him and what his conclusions are on those issues. It will only be in extreme circumstances such as those described by Clerk LJ in Gillies Ramsay, that the court will decline to enforce an otherwise valid adjudicator's decision because of inadequacy of the reasons given. The Complainant would need to show that the reasons were absent or unintelligible and that, as a result, he had suffered substantial prejudice.[29]*

> *In my opinion, a challenge to the intelligibility of stated reasons can succeed only if the reasons are so incoherent that it is impossible for the reasonable reader to make sense of them. In such a case, the decision is not supported by any reasons at all and on that account is invalid . . . .[30]*

> *It seems to me therefore that it is right that a recipient of a decision such as this should be entitled to know what it is the Adjudicator has decided and why.[31]*

Having considered the authorities, including the above extracts, Akenhead J in the case of *Balfour Beatty Engineering Services (HY) Ltd v Shepherd Construction Ltd*[32] drew a number of conclusions as to what is required by way of reasons if a decision is to be enforced. His observations include the following:-

> (f) *The reasons can be expressed simply. If the reasons are so incoherent that it is impossible for the reasonable reader to make sense of them, it will not be a reasoned decision.*
>
> (g) *Adjudicators are not to be judged too strictly, for instance by the standards of judges or arbitrators, in terms of the reasoning. This reflects the fact that decisions often have to be reached in a short period of time and adjudicators are often not legally qualified. It certainly reflects the fact that there has not been a full judicial or arbitral type process.*
>
> (h) *The fact that reasoning in a decision is repetitive, diffuse or even ambiguous does not mean that the decision is unreasoned.*

This is the only requirement contained in the Act or in the Code of Practice affecting the content of the adjudicator's decision. By way of contrast the New Zealand Act sets out at Section 45 a list of seven items being the only items adjudicators are entitled to consider in arriving at their decisions; Section 47 sets out the form of the adjudicator's determination in considerable detail and Section 48 sets out the substance thereof again in detail.

An article[33] by Australian authors reporting upon a recent case[34] before the Supreme Court of Western Australia suggests a greater readiness on the part of that court to quash an adjudicator's decision where the reasons given are inadequate or irrational:-

> This decision builds on a growing body of case law which emphasises that, whilst adjudications under the Act are intended to provide a rapid and informal means of resolving payment disputes, adjudicators must demonstrate, by their determinations, that they have engaged with all issues and adopted a rational approach to determining the dispute(s).

35 *The Adjudicator's decision shall allocate such fees, costs and expenses of the Adjudicator as he/she has authority to allocate under section 6(16) of the Act and under the provisions of this Code of Practice.*

This matter is discussed in more detail at page [ ] above. The reference to the Code of Practice is presumably to the provisions contained in paragraph 36.

36 *The fees, costs and expenses shall be reasonable in amount having regard to the amount in dispute, the complexity of the dispute, the time spent by the Adjudicator and other relevant circumstances.*

Presumably it would fall to the Court to decide whether or not such fees, costs and expenses are reasonable. It is thought that the fact that the parties may have agreed in advance an hourly rate or other structure for the assessment of the fees would be a factor, but not necessarily determinative, of the issue.

This paragraph must be read in conjunction with paragraphs 27 and 28 of the Code of Practice. The 2014 review conducted by the Glasgow Caledonian University Adjudication Reporting Centre[35] concluded "Results show an increasing tendency towards charging higher hourly rates, with around 29% in year 16 charging £176–£200 and 56% charging over £200 (per hour)".

The number of disputes with a value of less than £10,000 being referred to adjudication in the UK has diminished substantially over the years since 2001. This may be at least in part attributable to the cost of adjudication. Many would consider it simply not being worthwhile pursuing a debt of less than £5,000 or £6,000 through adjudication. There is probably a place in Ireland for adjudication schemes similar to arbitration schemes operated by the Chartered Institute of Arbitrators (Irish Branch). These enable inexperienced arbitrators gain familiarity with the process by conducting small arbitrations at a set fee. The fee could not be justified on the basis of the hours put into the exercise but can be justified in terms of the arbitrator / adjudicator gaining experience and reputation. Any such scheme would have to be on a voluntary basis and may be difficult to set up having regard to sub-section 2(5) of the Act, which prohibits opting out of the legislation.

The Regulatory Impact Analysis conducted for the Department of Public Expenditure and Reform in 2011 noted the rates charged by adjudicators in the

UK and went on to state *"Anecdotally, it is understood that equivalent rates in Ireland would be higher. It is understood that the hourly rates quoted in the reporting period would be similar to the level of fees paid for conciliation services in Ireland"*. Traditionally in Ireland arbitrators and conciliators have tended to charge fees at hourly rates similar to those charged by the larger firms of solicitors based in Dublin to their clients. These fees would normally be considerably higher than those charged by other professionals such as architects, engineers or quantity surveyors. Whilst London firms of solicitors would tend to charge at a higher rate than their Dublin counterparts, the same tendency for other professionals to charge at those rates has not occurred in the UK with the effect that the hourly rates charged by adjudicators there would not be as high as the hourly rates applied by conciliators in Ireland.

> 37   *Any document or information supplied for and/or disclosed in the course of the adjudication shall be kept confidential by the Adjudicator. He/she will only disclose such document or information if required to do so by law, or pursuant to an order of a court, or with the consent of all the parties to the payment dispute.*

There is no provision in the Act itself rendering adjudication a confidential process although this may well be implied by law, as it is in the UK. The confidentiality requirement imposed by this paragraph relates only to documents and information supplied or disclosed in the course of the adjudication. It may be inferred that a duty of confidentiality also arises in relation to any information provided prior to the nomination whether or not the appointment is accepted.

There is no protection afforded to the parties or to the adjudicator in respect of a subpoena *duces tecum* requiring adjudicators to produce documents provided to them in the course of the adjudication. The Scheme operable in the UK at regulation 18 of the Schedule provides as follows:

> The adjudicator and any party to the dispute shall not disclose to any other person any information or document provided to him in connection with the adjudication which the party supplying it has indicated is to be treated as confidential, except to the extent that it is necessary for the purposes of, or in connection with, the adjudication.

The first draft of the Code of Practice contained a similar provision, but this was omitted on further consideration. The existence of such a provision in the Scheme suggests that a party is entitled to provide the adjudicator with documents that are not disclosed to the other party. On the face of it, this would fly in the teeth of the basic principles of natural justice. If the adjudicator's decision was influenced or potentially influenced by any such confidential information, presumably it would be: 'necessary for the purposes of the adjudication' for the adjudicator to disclose the information. Otherwise there would be a clear breach of natural justice. Would every adjudicator be alert to this requirement? If an adjudicator was influenced by such information, how is the other party to know that this has occurred? It is submitted that having regard to the overarching requirements of constitutional justice, such a provision, had it been included in the code, would almost certainly be struck down by the courts in Ireland.

Neither the Act nor the Code of Practice contains any express provision requiring that any communication by one party to the adjudicator be provided also to the other party. It is submitted, however, that constitutional justice requires that this be done and an adjudicator, as with an arbitrator, must be vigilant to ensure that this requirement is observed.

The issue of confidentiality also arises in relation to whether documents prepared in connection with adjudications are privileged and therefore do not have to be disclosed[36]. In the case of *Walter Lilly v Mackay*[37], the question arose as to whether documents prepared by, or for, a claims consultant attracted legal professional privilege. Mr Justice Akenhead held that they did not.

However, it does appear that documents prepared for the purpose of adjudication do attract litigation privilege. In Australia the Supreme Court of Victoria held in *Dura (Australia) Constructions v Hire Boutique Living*[38] that such documents did attract litigation privilege. In the case of *A Struame (UK) Limited v Bradlor Developments Limited*[39] the court held that adjudication is a quasi-legal proceeding such as arbitration. It is thought that in Ireland, adjudication will also attract litigation privilege.

> 38 *The parties are responsible for their own legal and other costs incurred in connection with the adjudication in accordance with section 6(15) of the Act.*

This merely repeats sub-section 6(15) of the Act.

> 39 *The Chairperson may seek or, put in place arrangements to seek, details of adjudication cases from Adjudicators and which shall not include the names of the parties to a payment dispute. An Adjudicator, regardless of whether appointed to a payment dispute under section 6(3) or 6(4) of the Act, shall provide such anonymised information to the Construction Contracts Adjudication Service of the Department of Jobs, Enterprise and Innovation on each adjudication case within 21 days of the completion of the case. This will be used for the purpose of compiling statistical information relevant to adjudications conducted in accordance with the Act.*

The February 2016 draft contained a similar provision as set out in the first sentence but did not include any qualification by way of excluding the names of the parties to the dispute. It would seem appropriate that the chairperson would be in a position to fully investigate complaints made in relation to adjudicators and for this purpose to obtain full details of cases. It may also be appropriate for the chairperson to review the work of individual adjudicators on the chairperson's own initiative with a view to ensuring that adjudicators are conducting themselves in accordance with the Code of Practice and are producing decisions of a high standard.

The research carried out and the statistics published by institutions such as the Glasgow Caledonian University Adjudication Reporting Centre in the UK and the Adjudication Researching and Reporting Unit in New South Wales is of invaluable assistance in the consideration of any reforming legislation and to the industry generally. In the UK the available statistics only relate to appointments made by nominating bodies and not to appointments made by the agreement of the parties. Given that this provision applies to all adjudicators, whether appointed by the chairperson or not, the statistics and information available in Ireland should be complete and of particular interest.

References

1 [2009] EWHC 64 (TCC); 123 ConLR 15, [2009] Bus LR D76, [2009] All ER (D) 240 (Jan)
2 [2009] IEHC 465
3 [1980] IR 381
4 [1991] 3 WLR 1025; [1992] 1 QB 863, [1991] 3 All ER 211, [1991] 1 Lloyd's Rep 524, [1991] NLJR 343
5 [2008] EWHC 1836 (TCC); 120 ConLR 161, [2008] 3 EGLR 1, [2008] 44 EG 116, [2008] BLR 470, [2008] All ER (D) 378 (Jul)
6 [2014] EWHC 3710 (TCC); 157 ConLR 120, [2015] BLR 1
7 [2008] EWHC 3434 (TCC)
8 See page above
9 [2001] All ER (D) 138 (Apr); (2001) 80 ConLR 95, [2001] BLR 285, [2001] Lexis Citation 1085
10 [2001] All ER (D) 384 (Feb); (2001) 80 ConLR 14, [2001] BLR 207
11 The Scheme for Construction Contracts (England and Wales) Regulations 1998 at Schedule Part 1, para 12(a)
12 [2011] EWHC 269 (TCC)
13 *Locabail UK Limited v Bayfield Properties Limited* [2000] QB 451, para 25, [2001] All ER 65. [2000] 2 WLR 870, [2000] IRLR 96, (1999) Times, 19 November, 7 BHRC 583, [2000] 3 LRC 482,[1999] All ER (D) 1279
14 [1995] ILRM 408
15 Unreported, 23rd October 1997; [1997] Lexis Citation 4390
16 [2013] IEHC 402
17 [2000] 4 IR 412 at p441
18 [2007] IESC 42; [2008] 2 IR 40
19 [2010] EWHC 2344 (TCC); [2010] All ER (D) 82 (Oct)
20 [2011] EWHC 269 (TCC)
21 [1993] A.C. 646
22 [1997] 25 EHRR 577; [1997] ECHR 22299/93
23 [2008] EWHC 1836 (TCC); 120 ConLR 161, [2008] 3 EGLR 1, [2008] 44 EG 116, [2008] BLR 470, [2008] All ER (D) 378 (Jul)
24 2011] EWCA Civ 1617, [2012] Bus LR 1184, 141 ConLR 46, [2012] BLR 121, [2012] 13 Estates Gazette 92, [2011] All ER (D) 179 (Dec)

25 [2001] 1 WLR 700, [2001] ICR 564, (2001) Times, 2 February, [2000] All ER (D) 2425

26 [2008] QCA 83

27 Published by The School of Engineering & Built Environment, Glasgow Caledonian University, April 2012

28 [2008] EWHC 2025 (TCC); [2008] BLR 545, [2008] All ER (D) 190 (Oct)

29 Ramsay J: *Multiplex Construction UK Limited v West India Quay development Company* [2006] EWHC 1569 (TCC), 111 ConLR 33

30 *Gillies Ramsay Diamond v PJW Enterprises* [2004] BLR 131, 2004 SC 430, 2004 SLT 545, 2004 Scot (D) 3/1 para 31

31 *Thermal Energy Construction Ltd v AE & E Lentjes UK Ltd* [2009] EWHC 408 (TCC); [2009] All ER (D) 271 (Jan)

32 [2009] EWHC 2218 (TCC); 127 ConLR 110, [2009] NLJR 1475, [2009] All ER (D) 125 (Oct)

33 An article written for Lexology by Spencer Flay & Kristian Cywicki issued on 4th April 2016

34 *BGC Contracting Pty Limited v City Gate Properties Pty Limited* [2016] WASC 88

35 October 2014

36 For a full discussion in relation to the position in the UK see *The Role of Privilege in Adjudication*, Society of Construction Law Paper No 183 by Adrian Bell, April 2013 (www.scl.org.uk)

37 [2012] EWHC 649 (TCC), 141 ConLR 102, [2012] 20 LS Gaz R 25, [2012] BLR 249, [2012] 6 Costs LO 809, [2012] All ER (D) 32 (May)

38 [2011] VSC 477

39 [2000] BCC 333, (1999) Times, 29 June

13 Constitutional issues

Introduction

The courts have jealously guarded the right to natural justice when considering the law relating to adjudication in the UK. There is no doubt that the Irish courts would apply the principles applicable to natural justice with equal, if not greater, ardour. The principles of natural justice are guaranteed by the Irish Constitution (Article 40.3). That of itself does not necessarily give rise to any distinction between the law in Ireland and the UK in respect of adjudication, because the 1996 Act, as amended, does not exclude the application of natural justice in so far as that concept applies to the procedures governing the adjudication process (albeit paragraph 18 of the Scheme, which permits the adjudicator to receive confidential information from one party, may be an exception to this).

The question arises, however, as to whether the very essence of adjudication may offend against the concept of constitutional justice, i.e. the tenets of justice guaranteed by the Constitution, including natural justice. Does, for instance, the concept that a party may be obliged to make a payment on foot of an adjudicator's decision, which is patently incorrect, offend against the Constitution?

A separate question arises as to whether there are procedural rights guaranteed by the Constitution which would render the application of adjudication in Ireland inoperable within the timeframe contemplated. For instance, is the entitlement of a party to cross-examine, in circumstances where there is a dispute on an important issue of fact, so fundamental that a party to an adjudication in Ireland could render the process almost impossible of performance within the timeframe required, by insisting upon this right? These issues will be considered, amongst others, in this chapter. First, however, it is necessary to set out the constitutional background.

The relevant provisions in the Constitution

Article 37.1 of the Constitution provides as follows:

> *Nothing in this Constitution shall operate to invalidate the exercise of limited functions and powers of a judicial nature, in matters other than criminal matters, by any person or body of persons duly authorised by law to exercise*

such functions and powers, notwithstanding that such person or such body of persons is not a judge or a court appointed or established as such under this Constitution.

Article 40.3.1 provides:

The State guarantees in its laws to respect, and, as far as practicable, by its laws to defend and vindicate the personal rights of the citizen.

Article 40.3.2 provides:

The State shall, in particular, by its laws protect as best it may from unjust attack and, in the case of injustice done, vindicate the life, person, good name, and property rights of every citizen.

These are considered to be the main provisions of the Constitution relevant to the consideration of this topic. Other provisions and the preamble are also of some relevance.

It is Article 40.3.1 which guarantees fundamental rights. These fundamental rights include, but are not limited to, the right to fair and just procedures. In *Garvey v Ireland & Others*,[1] O'Higgins CJ observed:

The Constitution incorporates into our laws and their administration the requirements of natural justice, and by Article 40, s 3, there is guaranteed to every citizen whose rights may be affected by decisions taken by others the right to fair and just procedures. This means that under the Constitution powers cannot be exercised unjustly or unfairly. This applies as well to the Government as to any other authority within the State to which is given the power to take action which may infringe on the rights of others.

The concepts of constitutional justice and of natural justice are flexible and evolving. Natural justice is often stated as comprising the rule *audi alteram partem* and the rule *nemo judex in causa sua*, that is the entitlement to be heard and the rule that no one can be a judge in his own cause, respectively. These are very wide and flexible concepts. In fact, the common law concept of natural justice requires fair procedures be applied and the constitutional concept requires the same. Because both are flexible in terms of application to individual circumstances, it is certainly possible to argue that the distinction is illusory. For instance, the right to cross-examine, in certain circumstances, is considered by the courts in Ireland to be a constitutional right (*Re: Haughey*).[2] It is, however, also a right available in certain circumstances arising from the requirements of natural justice (*Galvin v The Chief Appeals Officer and the Minister for Social Welfare*).[3] The real distinction for the Irish courts arises from the fact that natural justice is guaranteed by the Constitution in Ireland, whereas in the UK it is not. The courts in the UK require that adjudicators comply with the requirements of natural justice, but there is an issue there as to whether the courts can insist upon compliance with natural justice where the legislation inherently permits or requires the adjudicator not to be so bound in certain circumstances. For instance, in the UK it does seem to be accepted by the courts that an adjudicator cannot be expected to hold an oral hearing, and there is no case where the court has held that an adjudicator was required to do so to meet the obligation to adhere to fair procedures. The Irish courts may have a difficulty with this approach.

There are no hard or fast rules as to what constitute fair procedures in relation to natural justice or constitutional justice. What might be a fair procedure in one context will not be a fair procedure in another. Matters which must be taken into account are as follows:

a   How relevant to the core issues is the matter in respect of which a right to cross examination, or other right, is sought?
b   How reliable are the documents if it is sought to rely on the documents rather than hear evidence?
c   What is at stake in terms of personal rights, e.g. is someone's reputation under attack?

There is nothing in principle to prevent the legislature from setting up a system of adjudication. The administration of justice is very largely administered through persons other than judges. Furthermore, the extent to which such persons are obliged to afford the parties an oral hearing, a right of representation, a right to cross-examine and other entitlements may depend on the nature of the case. In *Mooney v An Post*,[4] Keane J observed: 'the concept (of natural justice) is necessarily an imprecise one and what its application requires may differ significantly from case to case'.

Whether or not, therefore, an oral hearing is necessary, a right to cross examination should be extended or legal representation permitted may depend on the individual circumstances of each situation. It probably is the case that when put to the test, the Irish courts would find that there is no absolute right to any of these in respect of adjudication. Whether or not the entitlement is there would have to be judged by the circumstances of the individual case. Throughout this discussion it is worth bearing in mind the words of McMahon J in *Khan v HSE*:[5]

> The battle between fair procedures and efficiency has long since been fought and fair procedures have won out. The insistence on fair procedures governs all decision makers in public administration. It governs the courts as well. None of us can ignore the principle.[6]

In the case of *Construct Interiors NZ Limited v Peter William Jones and KMB Interior Contracts Limited*[7] the High Court of New Zealand was asked to rule upon an issue as to whether an adjudicator had complied with the requirements of natural justice. Serious allegations of fraud had been made by the executing party against the other party, and these were set out in an affidavit by the executing party, but the other party was not given an opportunity to respond. That other party satisfied the Court that it could have put forward evidence in rebuttal that might have persuaded the adjudicator that it had not fabricated documents as was claimed by the executing party. The executing party, KMB, argued that a number of submissions were made by both parties on the issue and that if the adjudicator were to keep affording the parties an opportunity to respond to the other's submissions, the whole process would be open ended. Cooper J found, however, as follows:

> (66) While it is no doubt correct, as Mr. Taylor submitted, that the Act is intended to provide for a fast and effective disputes resolution process so as to promote cashflow for work done on projects, it is clear that the Act requires observance of the principles of natural justice and there will be occasions when the quick resolution of the dispute will need to yield to that consideration. Where there has been a breach of natural justice and the procedure followed was not authorised by the Act, there can be no reason for the Court not to provide a remedy. Having regard to the seriousness of the allegation in the present case I consider there was such a breach. To adopt the language of Lord Diplock in the passage quoted at (41) above Re Erebus Royal Commission: Air New Zealand Limited v Mahon, SINZ were effectively deprived of the opportunity to adduce additional material of probative value which might have deterred the adjudicator from the finding that the documents had been fabricated. Not only was this important to the resolution of the substantive issues but it was crucial to the decision that was made as to the costs of the adjudication.

The New Zealand Court therefore came to a firm decision that where there is a conflict for the adjudicator between strict compliance with the legislation and the application of the principles of natural justice, the latter should prevail. However, the New Zealand Act expressly provides at Section 41(c) that the adjudicator must comply with the principles of natural justice. Neither the UK Act of 1996 nor the Scheme has a similar provision and the courts in the UK have tended to equivocate. In *Discain Project Services Limited v Opecprime Developments Limited*[8] Judge Bowsher said:

> At the same time, one has to recognise that the adjudicator is working under pressure of time and circumstance which make it extremely difficult to comply with the rules of natural justice in the manner of a court or an arbitrator. Repugnant as it may be to one's approach to judicial decision making, I think that the system created by the Act can only be made to work in practice if some breaches of the rules of natural justice which have no demonstrable consequence are disregarded.

This would infer that where a breach of the rules may have a material effect on the outcome, they cannot be disregarded. However, in a later decision by the same judge in a case between the same parties[9] the judge quoted, with approval, Judge Lloyd's judgment in *Glencot Development & Design Company Limited v Ben Barrett & Sons (Contractors) Limited*[10] to the effect that an adjudicator must observe the principles of natural justice: 'as fairly as the limitation imposed by Parliament permits'. This would infer that where a conflict arises, the principles of natural justice must yield to the requirements of the legislation. Therefore, if it was apparent that an issue affecting the decision could only be decided if an oral hearing was conducted, but the time restraints did not allow for such a hearing, the adjudicator must nonetheless give a decision without such a hearing. In such circumstances, it would appear that the adjudicator's decision would be enforced in the UK.

In practical terms, an adjudicator in such circumstances might request the parties to extend the time for his decision, so as to enable him to conduct an oral hearing on the issue and, if one party would not consent but the other would, to draw an inference from this against the party who withheld its consent. That, however, may itself raise an issue as to whether such an inference is fair. After all, the party withholding its consent is doing no more than insisting upon its statutory entitlement that the adjudicator make his decision within the time prescribed.

The Irish Code of Practice provides at paragraph 23 that adjudicators shall observe the principles of procedural fairness'. On this basis alone, the Irish courts may come to the conclusion that where a conflict arises between the time constraints imposed by the Act and the rules of natural justice, the former must yield to the latter. Where the issue is not of substantial materiality to the outcome of the adjudication, it is likely that the Irish courts will take a similar attitude to the courts in the UK to minor infringements of the rules of natural justice. Where, however, the breach has a significant effect on the outcome, it is likely that the courts would, on constitutional grounds, irrespective of the Code of Practice, come to the conclusion that the rules of natural justice are paramount and refuse to enforce the adjudicator's decision.

It is important to bear in mind what is being discussed here is an unavoidable conflict between the legislation and the rules of natural justice. The author is not aware of any case in the UK where this precise issue has arisen. The cases that have come before the courts have involved breaches that might have been avoided by the adjudicators and a decision nonetheless given within the time required by the legislation.

In practical terms, the issue of whether or not the requirements of natural justice can be met within the time constraints arises not infrequently for adjudicators at the outset of the adjudication. It may be obvious at the outset that it would be impossible to afford the parties the benefit of the principles of natural justice and at the same time render a decision within the time required. Guidance notes published in the UK[11] require in these circumstances that adjudicators refuse the appointment unless the parties are prepared to extend the time for their decision. This approach is consistent with the approach of the courts in the UK who have indicated that adjudicators should not proceed with an adjudication if they are convinced that they will not be able to do justice to the parties within the timeframe.[12]

The entitlement to an oral hearing/cross examination

The legal systems of common law countries are heavily weighted in favour of oral hearings. The general rule is that if there are disputes as to facts, a court or arbitrator will decide the issue on the basis of an oral hearing of the evidence. In construction adjudication, however, an oral hearing (as opposed to a procedural meeting) is very rare. This is simply because there is not sufficient time to conduct an oral hearing. Adjudicators therefore must simply do the best they can on the basis of the written submissions and documents put before them.

In *Re: Haughey*[13] the Dail Public Accounts Committee had put before it serious allegations by a senior member of An Garda Siochana of criminal conduct on the part of Mr Paraic Haughey. The Supreme Court confirmed, albeit these were not criminal proceedings, that Mr Haughey was entitled to a statement of the evidence in writing, that he should be afforded an opportunity to cross-examine his accuser and that he should be entitled to call rebutting evidence. The Law Reform Commission Consultation Paper on Hearsay in Civil and Criminal Cases 2010 stated:

> The decision in Haughey involves three important elements in the context of this consultation paper. First, the right to confront or to cross examination was specifically mentioned as a component of the right to fair procedures. Secondly, perhaps even more significantly, by deciding that the right to fair procedures was a constitutional right the Court held that legislation which attempted to prevent the ability to confront, including the legislation involved in the case itself could be constitutionally opened to doubt. Thirdly, the decision in Haughey was not limited to civil or criminal court proceedings but specifically involved any adjudicative processes where a person's rights are at issue.

In the adjudication process contemplated by the Act, a person's property rights will undoubtedly be threatened. Does this mean that a party to an adjudication under the Act will be entitled to insist, as a component of the right to fair procedures, that an oral hearing be held to determine issues of fact involving a conflict of evidence? If so, that could sound the death knell for adjudication, as a speedy outcome is the very essence of the process, and a decision made outside of the time limits provided will be invalid.

In almost any construction related dispute, a number of issues of fact require to be determined. By simply denying all of the facts relied upon by the claimant, an uncooperative respondent may be in a position to impair or even abort a relatively straightforward adjudication if that respondent is entitled to an oral hearing and a right of cross examination. This may not be insuperable in less complex disputes, but where the adjudication relates, say, to the final account and, therefore, usually, a multitude of issues, the process may be rendered inoperable.

In *Galvin v Chief Appeals Officer and the Minister for Social Welfare*[14] Costello J looked in some depth at the issue of whether natural justice required that an oral hearing involving evidence was required by the circumstances of the particular case. The applicant contended that he was entitled to certain benefits by way of pension by reason of having made qualifying contributions under the Social Welfare Acts in particular years. The first respondent drew certain inferences from the second respondent's records. The applicant maintained through correspondence that the conclusions drawn by the respondent were incorrect and requested an oral hearing. The legislation provided a discretion on the part of the first respondent as to whether he granted an oral hearing or not. Costello P held:

> The statute gives a discretion to the appeals officer to hold or not to hold an oral hearing. It is, of course, well established that where a decision-maker exercises statutory power in breach of the rules of natural / constitutional justice then the court will quash the decision. In this case, the issue is whether the rules of natural justice were breached in that fair procedures required that an oral hearing should have been held to determine the dispute which I have identified.

> There are no hard and fast rules to guide the appeals officer or, on an application for judicial review, this Court, as to when the dictates of fairness require the holding of an oral hearing. The case (like others) must be decided on the circumstances pertaining, the nature of the inquiry being undertaken by the decision-maker, the rules under which the decision-maker is acting, and the subject matter with which he is dealing and account should also be taken as to whether an oral hearing was requested. In this case there is no doubt that an important right was in issue (that is the applicant's right to a pension for life). The statute gives an express power to hold an oral hearing and to examine witnesses under oath; a request for an oral hearing was made. What I have to decide is (as Keane, J had to decide, in The State (Boyle) v The General Medical Services (Payments) Board [1981] ILRM 14) whether the dispute between the parties as to (a) the reliability of the evidence before the appeals officer, of the applicant and Mr Higgins on the one hand and (b) the accuracy of the departmental records on the other, made it imperative that the witnesses be examined (and if necessary cross-examined) under oath before the appeals officer.

> I have come to the conclusion that without an oral hearing it would be extremely difficult if not impossible to arrive at a true judgment on the issues which arose in this case.

Therefore, the court on the facts of that particular case came to the conclusion that the decision of the appeals officer could not be upheld.

The case of *J & E Davy t/a Davy, v Financial Services Ombudsman*[15] is particularly relevant to this issue. The Financial Services Ombudsman is entitled to adjudicate on disputes referred to it under the Central Bank Act 1942 as amended. The Act expressly provides that its object is to enable complaints to be dealt with in an informal and expeditious manner.[16] The Act also provides that the Ombudsman: 'when dealing with a particular complaint, is required to act in an informal manner and according to equity, good conscience and the substantial merits of the complaint without regard to technicality or legal form'.[17] Whilst the

Ombudsman is not required to hold an oral hearing, the Act confers on him the powers of a High Court Judge when hearing civil proceedings with respect to the examination of witnesses.

In the Davy case a firm of stockbrokers had given advice to a credit union in relation to certain investments. The credit union maintained that its committee should not be regarded as being experts and that it relied upon the expertise of Davy in relation to the investments. The Ombudsman arrived at his determination without an oral hearing and without providing certain documents that were before him to Davy. The High Court, and on appeal the Supreme Court, held that an oral hearing was essential for the purpose of constitutional justice. The Supreme Court at paragraph 111 stated (Finnegan J):

> I am satisfied that Section 57CE(5) empowers the respondent to proceed by way of examination and cross examination of witnesses where that is appropriate. The respondent may of course restrict the cross examination to those issues on which there is a conflict. Central to the respondent's decision was his finding on the expertise of the investment committee of the notice party. He formed this finding on the basis of witness statements which were not made available to the applicant. Fair procedures require that those officers of the notice party to whom the applicant gave oral advice should be produced for cross examination. Likewise in relation to the nature and suitability of the bonds, the expert who reported to the notice party and whose reports were before the respondent, although not furnished to the applicant, should be made available for cross examination.

Therefore the Ombudsman's decision in the matter was set aside by the court.

There are circumstances where there are important issues of fact in dispute, but the adjudicator reasonably believes that conducting an oral hearing would not assist in the resolution of that dispute. The onus is upon the party seeking the oral hearing to satisfy the adjudicator (or Ombudsman as the case may be) that there is an alternative construction to be gleaned from the oral evidence that may affect the outcome of the adjudication. In the *Davy* case, Davy sought to establish through oral evidence that they had carefully explained the nature of the investment to the credit union. If they were correct in this, that might well have influenced the decision of the Ombudsman. In the case of *Star Homes (Middleton) Limited v The Pensions Ombudsman*[18] the applicant did not explain to the Ombudsman, or indeed to the court, what it was that it had sought to establish through the oral evidence. The court reached the conclusion that the Ombudsman would not be acting unreasonably or irrationally (that being the test) in concluding that an oral hearing would not affect the outcome and that it should therefore be denied. It should be borne in mind, however, that the facts in the *Star Homes* case were unusual in that the oral evidence would have related to an alleged event involving the applicant and a deceased employee. The oral hearing by necessity, therefore, would have been a rather one-sided account of the event, which might equally have been put forward in written submission.

In *Lyons & Another v Financial Services Ombudsman*[19] Hogan, J reviewed a number of cases where the Ombudsman had refused to grant an oral hearing, and his decision was upheld by the Court. It is of interest to note that the court took account of the purpose of the legislation and the practicality of allowing oral hearings. The court observed at paragraph 21 of the judgment:

> All of this brings us to the nub of the present appeal, namely, the decision of the FSO to reject the Appellant's complaints without an oral hearing. It goes without saying in the context of an adjudicatory system which is statutorily designed to be informal and expeditious that the courts should be reluctant to impose some form of adversarial court-style model . . . Indeed, as MacMenamin J pointed out in Ryan,[20] if such a model were in fact to be imposed on the Ombudsman, it would mean in reality that the office simply could not function. The FSO cannot be regarded as some form of miniature version of the Commercial Court and, as counsel for the Ombudsman, Mr McDermott, submitted, it could not practically function if this is what was expected of it.

It may be inferred from this and from the various decisions referred to in that judgment[21] that the courts will be less ready to infer an obligation to have an oral hearing where the very purpose of the statutory adjudication is to provide an expeditious process, than might otherwise be the case. This, however, is always a matter of balance, and this point is only one of many to be taken account of in having regard to the entitlement of the parties to the benefit of constitutional justice.

Having reviewed the judgments of the Court upholding the decisions of the Ombudsman not to grant an oral hearing, however, in the particular case Hogan J found that an oral hearing was essential, because: 'The appellants could not realistically hope to establish the underlying merits of their case without an oral hearing'.

This view was reinforced by the same judge in *O'Neill v Financial Services Ombudsman*[22] where it was stated in the context of the constitutional guarantee of basic fairness of procedures:

> Where, as here, the conflict of fact is a stark one and the resolution of this conflict is central to the fair disposition of the complaint, then it is impossible to avoid the conclusion that some form of oral hearing is objectively necessary in order to give effect to this constitutional guarantee'.

The Act does not specify whether or not the adjudicator may hold oral hearings. It merely requires the adjudicator to reach a decision within the specified times. The Code of Practice requires the adjudicator to observe the principles of procedural fairness and, at paragraph 24, allows the adjudicator, inter alia, to invite written submissions and evidence from both parties and to meet jointly with, and question, the parties and their representatives. Under paragraph 27 of the Code of Practice, the adjudicator is obliged to be mindful of the fact that an oral hearing may be required.The question is whether the adjudicator must hold an oral hearing where, in the words of Costello P, as quoted above: 'without an oral hearing it would be extremely difficult if not impossible to arrive at a true judgment on the issues which arose in this case'. That quote has to be taken within the context of the particular case. If the issues which arise are not particularly germane or important, it is submitted the necessity for an oral hearing would not arise. Where, however, the issues go to the heart of the matter, it may be necessary, in order to meet the obligations of fair procedures imposed by the Constitution, to conduct an oral hearing on the particular issue allowing for the right of cross examination. It may be that an adjudicator would not be faulted for not doing so, unless an oral hearing is requested. Where, however, an oral hearing is requested and the issue is of importance, if not crucial, case law would suggest that an adjudicator would be obliged to grant an oral hearing. Potentially, that can give rise to major logistical and jurisdictional issues, because it may not be possible, due to the absence of witnesses or otherwise, to conduct the oral hearing within the time for the adjudicator to make his decision.

It is of interest to note that the issue of whether or not an adjudicator should conduct an oral hearing is not even considered in the guidance notes published by the Adjudication Society and Chartered Institute of Arbitrators in the UK.[23] The guidance notes warn of the various pitfalls whereby an adjudicator's decision might be challenged on the grounds of natural justice, but do not contemplate even in that context that failure to conduct an oral hearing would in any circumstances give rise to a breach of natural justice. At paragraph 3.17 the guidance document notes that the rules of natural justice apply, but goes on to state that the duty to apply procedural fairness is: 'qualified due to: (1) the constraints inherent in the tight timescales under which the legislation expects the adjudicator to conduct the adjudication; and (2) the provision and nature of the adjudicator's decision'.

Right of access to the courts

Article 40.3.1 of the Constitution by inference guarantees to citizens rights of access to the courts. However, Article 37.1, as quoted above on page 170 limits that right of access where proceedings are not of a criminal nature. It is well established that tribunals and other bodies set up by statute or by the agreement of individuals may exercise judicial powers. Thus, in *Re: Haughey*[24] Henchy J at page 10 stated:

> As to the Act itself, even if there be validity in the submission that the Act of 1970 empowers the Committee to exercise functions and powers of a judicial nature, since such functions and powers are limited and are not exercisable in a criminal matter they are validated by Article 37 of the Constitution.

Furthermore, the Courts will not normally interfere with the decisions of such bodies provided they are acting within jurisdiction. For example, in *Galvin v The Chief Appeals Officer and the Minister for Social Welfare*,[25] Costello P found at page 9:

The issue was whether contributions were paid and on the one hand the applicant claimed that they had been and on the other there was no departmental records of payments having been received. There was no evidence on which the deciding officer could infer why no contributions were made. But this error does not entitle the applicant to an order of certiorari as it was made within jurisdiction.

Similarly An Bord Pleanala in exercising its functions under the Planning Acts makes very important decisions affecting the property rights of citizens with no right of appeal against that decision, provided the board in hearing the appeal adheres to the principles of natural justice and acts within jurisdiction.

These are just a couple of examples whereby bodies outside of the courts' system are empowered to make judicial decisions that are binding upon the parties. There are many others. In *Daly v Revenue Commissioners*[26] Costello P said:

> But legislative interference in property rights occurs every day of the week and no constitutional impropriety is involved when, as in this case, an applicant claims that his constitutionally protected right to private property referred to in Article 40.3.2 has been infringed and that the State has failed in its obligation imposed on it by that Article to protect his property rights, he has to show that those rights have been subject to 'an unjust attack'. He can do this by showing that the law which has restricted the exercise of his rights or otherwise infringed them has failed to pass a proportionality test.

The fact, therefore, that an adjudicator may be appointed under a statutory provision to make decisions of fact that may be binding provided they meet the requirements of natural justice and are within jurisdiction, is not, on the face of it, likely to offend against the Constitution. On the contrary, one might argue that given the adjudicator's decision may be challenged through litigation or arbitration, adjudication is far less an attack on the property rights of the citizen than might be lawfully sustained, for example, at the hands of An Bord Pleanala, without a right of appeal. What makes adjudication unique in terms of constitutional consideration is the combination of:

a   the possibility/likelihood of rough justice or injustice occurring because of the time restraints;
b   the possibility of it being impossible to fully apply fair procedures within those time constraints;
c   the fact that an unsuccessful party has to pay the successful party pending the outcome of litigation or arbitration and the possibility that the monies may not be recovered even if the paying party is successful in such litigation or arbitration.

If any one of these three elements were missing from the system of adjudication, it is unlikely that any constitutional issues would arise. The question is whether the combination of all three taken together renders the legislation incapable of performance in accordance with constitutional justice. Mr Justice Frank Clarke of the Supreme Court touched on some of these issues in a paper on the Irish legislation delivered to Engineers Ireland in January 2014.[27] He reached a number of conclusions:

1 Notwithstanding that an adjudicator's decision is not finally binding, the process nonetheless must be conducted in accordance with constitutional justice.[28]
2 The need to comply with fair process overrides the practicality issues, which arise from the requirement to provide a speedy resolution.[29]
3 Notwithstanding the time constraints adjudicators are likely to be required to provide any information on which they rely to the parties for comment.[30]

At page 10 of the paper the judge raised the question of what adjudicators are to do in circumstances where they believe that they cannot conduct the adjudication in compliance with the requirements of constitutional justice within the time constraints required by the legislation. In response, he raised by way of analogy the approach the courts have taken to the very strict time requirements in respect of companies placed in examinership. Unless an acceptable scheme for the company's survival is placed before the court within 70 days (or in certain circumstances 100 days), the company will be put into liquidation. In the case of *O'Brien's Irish Sandwich Bars Limited v Companies Acts*[31] the court took the view that it did not have sufficient time to conduct a number of hearings that would be necessary to allow different landlords of the company to argue against the manner in which it was proposed to treat their leases in the scheme of arrangement. In circumstances where the court could not conduct fair hearings in the time available, the court felt obliged to reject the scheme. The implication is that an adjudicator may well be found to have the same obligation, i.e. to refuse to render a decision where it is impossible to meet the requirements of natural or constitutional justice within the time allowed.

The presumption of constitutionality

Henchy J in the Supreme Court decision in *Re: Haughey*[32] stated the extent of the presumption of constitutionality concisely as follows:

> It is a well established rule of interpretation that statutes which have been enacted after the coming into operation of the Constitution enjoy a presumption of constitutionality and 'one practical effect of this presumption is that if in respect of any provision or provisions of the Act two or more constructions are reasonably open, one of which is constitutional and the other or others are unconstitutional, it must be presumed that the Oireachtas intended only the constitutional construction and a Court called upon to adjudicate upon the constitutionality of the statutory provision should uphold the constitutional construction. It is only when there is no construction reasonably open, which is not repugnant to the Constitution that the provision should be held to be repugnant' (see McDonald v Bord na gCon [1965] IR 217) Furthermore, 'an Act of the

Oireachtas, or any provision thereof, will not be declared invalid where it is possible to construe it in accordance with the Constitution; and it is not only a question of preferring a constitutional construction to one which would be unconstitutional where they both may appear to be open but it also means that an interpretation favouring the validity of the Act should be given in cases of doubt', see East Donegal Co-Operative v Attorney General [1970] IR 317.

(page 227)

In the same judgment at page 229 Henchy J stated:

As was stated by the Supreme Court in East Donegal Co-operative v Attorney General [1970] IR 317 'the presumption of constitutionality carries with it not only the presumption that the constitutional interpretation or construction is the one intended by the Oireachtas but also that the Oireachtas intended that proceedings, procedures, discretions and adjudications which are permitted, provided for, or prescribed by an Act of the Oireachtas are to be conducted in accordance with the principles of constitutional justice. In such a case any departure from those principles would be restrained and corrected by the Courts'.

The important point in the context of the constitutionality of the Act is that, where legislation is capable of being interpreted so that it is constitutionally compliant or not, the court will presume that the legislature intended the compliant interpretation to apply. Subsection 6(1) of the Act confers the entitlement to adjudication on a party to a construction contract in respect of 'any dispute relating to payment'. This expression is capable of a narrow and wide interpretation. It may be that by applying a narrow definition so that the adjudicator is simply valuing the works done at a particular date and no more, the court could readily justify the legislation in terms of its constitutionality. Almost any dispute under a construction contract can be described as 'relating to payment'. If the Act was interpreted as permitting adjudication in relation to very complicated issues such as payment for extensions of time, the potential for injustice through mistakes and/or lack of a fair hearing would be potentially far greater. It is not clear from the debates in the Dail or the Senate why it was decided to describe the nature of disputes in this fashion as opposed to the broader wording used in the 1996 Act, which simply relates to 'disputes'. The regulatory impact analysis is also silent on this issue. It may be that the intention of this qualification was to protect the legislation against constitutional challenge. A narrow definition of the expression potentially overcomes the broader attack on the basis that any legislation that contemplates payment on foot of incorrect decisions must be unconstitutional. It also affords protection against constitutional attack focussed on the test of proportionality and necessity.

The legislation would be seen as perhaps going no further than was necessary to ensure that the executing party was protected against the paying party in refusing to make payment unjustly.

However, it has to be said that, on the basis of international precedent, the words: 'relating to payment' should be broadly interpreted to include disputes as to extension of time entitlement and latent defects provided, as they invariably do, the disputes have monetary consequences (see page 52 above).

The justification for the legislation

No doubt when the constitutionality of the Act is tested before the courts, there will be two main points to the arguments in support of the legislation. The first is an issue of balance between the constitutional rights of the paying party as against the constitutional rights of the party seeking payment. In the majority of cases in the UK, the party seeking relief is a sub-contractor. The purpose of the Irish legislation was primarily to protect sub-contractors from abusive behaviour by main contractors through a system providing for the regular payment for the work to the sub-contractor as the project progresses. It will be argued that the limited infringement into the property rights of the paying party is justified in terms of meeting the legitimate expectations of the executing party, which expectations were not being met prior to the introduction of the legislation.

Those defending the legislation will also argue that constitutional issues do not arise, because the decision of the adjudicator is not final and the binding nature of his decision gives rise to only a temporary situation, pending the outcome of the dispute being litigated. These two arguments are not necessarily separate. The first gives rise to questions of proportionality, and the temporary nature of the payment is relevant to those questions.

The strongest argument in favour of adjudication is a practical one which may not readily be appreciated by the courts in Ireland. More than in any other industry disputes are commonplace in the construction industry. Disputes relating to the value of construction works, extensions of time, defects etc. arise in relation to almost every project. Most disputes are resolved without recourse to the dispute resolution mechanism of the contract but many are not. Conciliation and mediation have a high success rate but twenty or thirty percent of these are unsuccessful. As a result major contractors find themselves in arbitration as part and parcel of the process on a regular basis. The truth is that even the major players cannot afford the cost of arbitration, either financially or in terms of the disruption caused to their businesses, on a regular basis and smaller contractors or sub-contractors may have to often settle for less than their fair entitlements by reason of that cost. Given that disputes requiring of an authorative decision are such a regular feature in the industry, some relatively inexpensive and swift resolution process is required. Adjudication is the answer to that dilemma. The judges of the TCC Court fully understand this because appointments are made to that court from the pool of barristers specialising in construction law. Barristers appointed to the bench in Ireland do not have that specialist appreciation of the problem. It is submitted however that this background is a significant factor to be taken account of by the court, particularly in the context of proportionality.

The issue of proportionality

In *Heaney v Ireland*,[33] Costelloe P proposed the following test be applied when considering whether a legislative provision affecting fundamental rights offends against the Constitution:

(a) The objective of the impugned provision must be of sufficient importance to warrant overriding a constitutionally protected right, and must relate to concerns, pressing and substantial, in a free and democratic society.
(b) The means chosen must pass a proportionality test. They must impair the right as little as possible; and must be such that their effect on the right was proportional to the objective.[34]

The balance in Heaney was one involving the common good in terms of the administration of justice and not the property rights of an individual. The balance arising from the Act is one between the property rights of the paying party and the property rights of the executing party. The reference, therefore, to: 'a free and democratic society' is not perhaps relevant. Otherwise, however, the test would appear to be apt.

Obviously, arguments can be made both for and against the legislation by reference to both aspects of the test. One can readily see, however, how the interpretation of the expression: 'any dispute relating to payment' (subsection 6(1)) might be very relevant to both aspects.

The temporary nature of adjudication

Of course, it may be that the court would have no difficulty in robustly supporting the legislation simply on the grounds that the adjudicator's decision is merely a temporary measure. It might be said that most construction contracts include provisions whereby the certifier's decision is binding on the parties pending litigation or other means of dispute resolution. In so far as the purpose of the exercise is to ensure that the executing party is not unjustly denied his cash flow, adjudication is an improvement on that situation, because the adjudicator will be wholly independent of the parties whereas the certifier is invariably appointed by the employer under the main contract and, in relation to sub-contracts, is often the main contractor. Undoubtedly, adjudication is an improvement in this regard and to that extent it would be very regrettable if the legislation failed for constitutional reasons. Nonetheless, the current situation arises out of contracts freely entered into by the parties, whilst the proposed system of adjudication is being imposed by legislation. In the absence of unfair terms of contract legislation applying as between executing and paying parties within the construction industry (which some might say is long overdue), the parties are contractually free to agree such terms as they like. The legislature, however, is not entitled to pass such legislation as it likes. Even where, as here, the legislation is well intentioned and deservedly enjoys widespread support within the industry, it still must have regard to the constitutional rights of the paying party. Therefore, the comparison with the role of the certifier, and the fact that the certifier's decision may be enforced irrespective of whether it is right or wrong by the courts if the contract so provides, is not a valid comparison. The question here is whether the legislature is entitled to introduce legislation whereby the paying party may lose substantial money by operation of law, notwithstanding that the adjudicator's decision is found to be wrong.

A decision of the District Court or of the Circuit Court is not binding on the parties pending the outcome of an appeal to a higher court. In theory, a decision of the High Court is binding, but in practice, a stay on payment is invariably given by the court pending the outcome of the appeal. It can be said, therefore, that the decision of an adjudicator is rather unique in being immediately enforceable, notwithstanding that it may be obviously incorrect or found at a later date to be incorrect through litigation or other means of dispute resolution. Would legislation, which required in civil proceedings that money found due by a District Court Judge or a Circuit Court Judge had to be paid over by the losing party notwithstanding the fact of an appeal, be found by the courts to be constitutional? If a decision of a judge is not temporarily binding why should the decision of a lay person be temporarily binding?

Suppose the Act provided for a quick means of obtaining a judgment of the District Court rather than an adjudicator's decision, and the judgment was binding pending any appeal. It is submitted that this may well be found to be compliant with the Constitution. However, inherent in such a system would be some form of judicial hearing before a competent judge appointed under the Constitution. Inherent in the adjudication process is the absence of a hearing and the decision of a person who is not a judge. Is the binding nature of the adjudication process in these circumstances to be regarded as an unjust attack on the property rights of the paying party?

It is submitted that it is unlikely a court would come to the conclusion that the Act cannot be challenged on constitutional grounds, simply because the decision is a temporary one, albeit binding in the interim. If that is correct, then the issue becomes one of whether a requirement to protect the interests of the executing party is sufficient to justify the risk caused to the fundamental rights of the paying party and whether the terms of the legislation are sufficiently proportionate to meet the balance of the requirement.

It seems unlikely that a court would reach the conclusion the legislation is incapable of being operated in accordance with constitutional justice in principle, and that it must therefore be struck down. Where the input of the court is likely to be more significant in an Irish context than elsewhere is with reference to decisions that are patently incorrect. It is unlikely that an Irish court will feel the same sense of obligation, as have the courts elsewhere, to uphold an adjudicator's decision, which is clearly incorrect on the face of the document. The courts are unlikely to follow the precedent set by the UK courts in cases such as *Bouygues UK Ltd v Dahl-Jensen UK Ltd.*[35] It may also be anticipated that the Irish courts would require an adjudicator to hear evidence and allow cross examination on any important issue of fact in dispute. The courts are of course in a position to exercise a certain amount of control over the risk of injustice through their inherent jurisdiction to refuse enforcement of an adjudicator's decision. It may be that the courts would use that discretion somewhat more widely than it is availed of in the UK. The courts might for instance take the view that a decision by the adjudicator not to allow a hearing on a particular issue or issues was justified in the context of the relevant time limits but nonetheless put a stay on the enforcement of the decision, in whole or in part, if the court was of the view that this rendered the decision less safe than might otherwise be the case. Edwards-Stuart J. in *Galliford Try v Estura*[36] put a stay on £2.5million in circumstances where the contractor through a *"smash and grab"*[37] adjudication was entitled to payment of £4million and the employer could not afford to pay that sum and also challenge the amount of the decision through arbitration or litigation. This was seen in the UK as an exception to the normal rules whereby adjudicators decisions are enforceable but it may be the case in Ireland that the courts would avail more frequently of their jurisdiction to grant a stay to avoid injustices.

It is submitted that, constitutionally, this legislation skates on thin ice. It undoubtedly puts the paying party at the risk of injustice without redress. No doubt the legislature was aware of this, but thought the legislation to be of sufficient importance to justify the risk. It is perhaps tangentially surprising that, if the legislature considered the legislation to be so important, it did not go to the trouble of ensuring a system for the immediate enforcement of adjudicators' decisions by the courts. Without a robust procedure for enforcement being put in place the effectiveness of the adjudication process is put in jeopardy.

Discretion as to enforcement

The Scottish case of *Whyte & Mackay Limited v Blyth & Blyth Consulting Engineers Limited*[38] may offer the Irish courts an interesting precedent as to how a balance is to be achieved between the public interest aspect of the legislation on the one hand, and the constitutional rights of parties on the other. In that case, the Outer House of the Court of Session refused to enforce an adjudicator's decision where it found that it was unnecessary to do so, and indeed unfair to do so, and in circumstances where the adjudicator's decision did not relate to the public interest issues, which the legislation was designed to protect.

The defendants, Blyth & Blyth, were a firm of consulting engineers. The plaintiffs held a premises under long lease. Blyth & Blyth had given structural advice in relation to construction works on the premises, which were said to be negligent. The unusual feature to the case was that the actual repair works would not have to be carried out until the property was due to be handed back to the freeholder in 2036. The adjudicator found that the defendants were negligent and made a determination that the defendants pay to the plaintiffs £3 million. A number of aspects concerned the judge, Lord Malcolm:

- As this was a complex case involving professional negligence, adjudication was thought to be unnecessary and inappropriate.
- This would be particularly so having regard to the fact that the plaintiffs would not need the money for another twenty years or so to carry out the repairs.
- The adjudicator was not a lawyer and was given an extremely difficult task of making a determination within a very tight timeframe. It was clear that the adjudicator was not familiar with the legal concepts and was uncomfortable in arriving at his decision.

None of these concerns, however, would justify the court in not enforcing the adjudicator's decision on the basis of precedent. The legislation permits adjudication to take place at any time and the plaintiff/claimant was therefore entitled to seek redress immediately, notwithstanding there was no immediate loss. The defendant, however, successfully relied upon Article 1 of the first protocol to the Convention of the Human Rights Act 1998, which incorporated the 1950 European Convention for the Protection of Human Rights and Fundamental Freedoms. This Article provides:

> Every natural or legal person is entitled to the peaceful enjoyment of his possessions. No one shall be deprived of his possessions except in the public interest and subject to the conditions provided for by law and by the general principles of international law.
>
> The preceding provisions shall not, however, in any way impair the right of a State to enforce such laws as it deems necessary to control the use of property in accordance with the general interest or to secure the payment of taxes or other contributions or penalties.

The Court found that in the circumstances of the particular case the entitlement of Blyth & Blyth to the peaceful enjoyment of its possessions took precedence over the public interest considerations, because the adjudication legislation was introduced to protect cash flow as its primary objective, and that consideration did not apply to the circumstances of the particular case.

One might have difficulty in distinguishing the principles set out in Article 1 of the first protocol from similar considerations set out in the Irish Constitution.

If the Irish courts were to follow a similar route, they may come to the conclusion that whether or not an adjudicator's decision should be enforced would depend upon making a balance of this nature. If the detriment to the respondent is disproportionate to the public interest factors, the court may not enforce the decision.

In an interesting article on this case, Andrew Bartlett QC points out that the case was unreported in the Building Law Reports and Construction Law Reports and may be considered by some as a: 'Scottish aberration, best not publicised'.[39] The author concludes that the judgment, far from being an aberration: 'fits comfortably into a correct understanding of the nature and purpose of adjudication'.

References

1 [1981] IR 75
2 [1971] IR 217
3 [1997] 3 IR 240
4 [1994] ELR 103
5 [2008] IEHC 234 at page 9
6 Approved by Hardiman J of the Supreme Court in *Dellway Investments Limited v NAMA* [2011] 4 IR 1 at 280; [2011] IESC 13
7 Civ 2010-404-897
8 [2000] BLR 402

 9 (2001) 80 ConLR 95, [2001] BLR 285, [2001] Lexis Citation 1085, [2001] All ER (D)
    138 (Apr)
10 [2001] All ER (D) 384 (Feb); (2001) 80 ConLR 14, [2001] BLR 207,
11 *Guidance Note: Jurisdiction of the UK Construction Adjudicator*, published by the
    Adjudication Society and the Chartered Institute of Arbitrators, third edition, paragraph
    2.38
12 *Dorchester Hotel Limited v Visit Interiors Limited* [2009] BLR 135; [2009] EWHC 70
    (TCC); [2009] Bus LR 1026, 122 ConLR 55, [2009] All ER (D) 264 (Feb)
13 [1971] IR 217
14 [1997] 3 IR 240
15 [2010] 3 IR 324
16 Section 57BF(1)
17 Section 57BK(4)
18 [2010] IEHC 463
19 [2011] IEHC 454
20 *Ryan v Financial Services Ombudsman*, High Court, 23 September 2011
21 *Molloy v Financial Services Ombudsman*, High Court, 15 April 2011; *Cagney v
    Financial Services Ombudsman*, High Court, 25 February 2011; *Caffrey v Financial
    Services Ombudsman* [2011] IEHC 285
22 [2014] IEHC 282
23 *Guidance Note: Jurisdiction of the UK Construction Adjudicator*, third edition, pub-
    lished by the Adjudication Society and Chartered Institute of Arbitrators (July 2015)
24 [1971] IR 217
25 [1997] 3 IR 240
26 [1995] 3 LR 1
27 Adjudication – The Role of the Courts, Mr Justice Frank Clarke, 29 January 2014
28 Page 5: 'I do not think that it can be argued that constitutional justice does not apply at
    all for a number of reasons'
29 Page 11
30 Page 8
31 [2009] IEHC 465
32 [1971] IR 217
33 [1994] 3 IR 593 at 607
34 *Chawke v Orr*, [1990] 3 SCR 1303, 1335–1336
35 [1999] All ER (D) 1281; (1999) 70 ConLR 41, [2000] BLR 49, [1999] Lexis Citation
    3672, – See page 94 above
36 [2015] EWHC 412 (TCC); 159 ConLR 10, [2015] BLR 321, [2015] All ER (D) 01
    (Mar)
37 The expression *"smash and grab"* has been used regularly to describe an entitlement
    arising from the paying party's failure to serve a pay less notice in response to an appli-
    cation for payment under the 2009 Act
38 [2013] Scot (D) 5/4; [2013] CSOH 54, 2013 SLT 555
39 *The Limits of Adjudication: The Impact of the European Convention on Human Rights*,
    an SCL Paper, published December 2014

14 Miscellaneous matters

Section 11: Expenses

The expenses incurred by the Minister in the administration of this Act shall be paid out of moneys provided to the Oireachtas.

Does this mean that the Minister cannot charge a fee for the appointment of an adjudicator? In other countries providing for appointments being made by a State appointed body, that body is authorised to charge a fee for making the appointment.

Regulation of adjudicators' fees

Legislation in certain countries, usually through regulations rather than the primary legislation, regulates the fees payable to adjudicators. For instance, in Singapore the fee payable to an adjudicator cannot exceed SGD2,000 per day or SGD250 per hour with an overall cap of SGD2,000 where the claim does not exceed SGD20,000 and of 10 per cent of the sum claimed in other cases.[1]

The New South Wales Adjudication Research and Reporting Unit Annual Report for the year 2012/2013 indicates that the average fees charged by adjudicators was only AUD2,100, while the fees payable to nominating bodies averaged about AUD900. More than half of the claims involved sums of less than AUD25,000.

Liability of the employer/owner to a sub-contractor

The legislation in some jurisdictions establishes a potential entitlement for a sub-contractor to be paid directly by the employer where the main contractor defaults. For instance, under the New Zealand Act a claimant under a commercial construction contract may seek through the adjudication liberty to register a charging order against the title of the owner of the property where the main contractor and the owner are associated.[2] This relationship often materialises in an Irish context where the developer owns the land in the name of one company and carries out the construction works in the name of another company in the same group. In such circumstances, in New Zealand, the adjudicator may determine that the owner is jointly and severally liable with the respondent construction company.

References

1   Building and Construction Industry Security of Payment Regulations 2005, paragraph 12(b)
2   Sections 29 and 49

Appendix A

Construction Contracts Act 2013

ARRANGEMENT OF SECTIONS

Section

SCHEDULE

Provisions to apply to matters regarding payments

Acts Referred to

CONSTRUCTION CONTRACTS ACT 2013

An act to regulate payments under construction contracts and to provide for related matters.

[*29th July*, 2013]

BE IT ENACTED BY THE OIREACHTAS AS FOLLOWS:

1.—(1) In this Act—

"construction contract" means (subject to *subsection (2)* and *section 2*) an agreement (whether or not in writing) between an executing party and another party, where the executing party is engaged for any one or more of the following activities:

(*a*) carrying out construction operations by the executing party;

(*b*) arranging for the carrying out of construction operations by one or more other persons, whether under subcontract to the executing party or otherwise;

(*c*) providing the executing party's own labour, or the labour of others, for the carrying out of construction operations;

"construction operations" means, subject to *subsections (3)* and *(4)*, any activity associated with construction, including operations of any one or more of the following descriptions:

(*a*) construction, alteration, repair, maintenance, extension, demolition or dismantling of buildings, or structures forming, or to form, part of the land (whether permanent or not);

(*b*) construction, alteration, repair, maintenance, extension, demolition or dismantling of works forming, or to form, part of the land, including (without prejudice to the foregoing) walls, roadworks, power-lines, telecommunications apparatus, aircraft runways, docks and harbours, railways, inland waterways, pipe-lines, reservoirs, water-mains, wells, sewers, industrial plant and installations for purposes of land drainage, coast protection or defence;

(*c*) installation in any building or structure of fittings forming part of the land, including (without prejudice to the foregoing) systems of heating, lighting, air-conditioning, thermal insulation, ventilation, power supply, drainage, sanitation, water supply or fire protection, or security or communications systems;

(*d*) external or internal cleaning of buildings and structures, so far as carried out in the course of their construction, alteration, repair, extension or restoration;

(*e*) operations which form an integral part of, or are preparatory to, or are for rendering complete, such operations as are previously described in this subsection, including site clearance, earth-moving, excavation, tunnelling and boring, laying of foundations, erection, maintenance or dismantling of scaffolding, site restoration, landscaping and the provision of roadways and other access works and traffic management;

(*f*) painting or decorating the internal or external surfaces of any building or structure;

(*g*) making, installing or repairing sculptures, murals and other artistic works that are attached to real property;

"executing party", in relation to a construction contract, means—

(*a*) where the parties to the construction contract are a contractor and the person for whom the contractor is doing work under the contract, the contractor, or

(*b*) where the parties to the construction contract are a contractor and a subcontractor or are 2 subcontractors, the subcontractor or whichever of the subcontractors agrees to execute work under the contract;

"main contract" means a construction contract such as is referred to in *paragraph (a)* of the definition of "executing party";

"Minister" means the Minister for Public Expenditure and Reform;

"other party", in relation to a construction contract, means the party to the construction contract who is not the executing party;

"payment claim" means a claim to be paid an amount under a construction contract;

"payment claim date", in relation to a construction contract, means the date when a payment claim in relation to an amount due under the construction contract is required to be made;

"payment claim notice" has the meaning assigned to it by *section 4*;

"payment dispute" has the meaning assigned to it by *section 6*;

"subcontract" means a construction contract such as is referred to in *paragraph (b)* of the definition of "executing party";

"subcontractor" means a person to whom the execution of work under a construction contract is subcontracted by the contractor or another subcontractor; "work", in relation to a construction contract, means any act done in furtherance of the construction contract under the terms of the construction contract.

(2) In this Act references to a construction contract include an agreement, in relation to construction operations, to do work or provide services ancillary to the construction contract such as—

(*a*) architectural, design, archaeological or surveying work,

(*b*) engineering or project management services, or

(*c*) advice on building, engineering, interior or exterior decoration or on the laying out of landscape.

(3) Subject to *subsection (4)* references in this Act to construction operations do not include the manufacture or delivery to a construction site of—

    (*a*) building or engineering components or equipment,

    (*b*) materials, plant or machinery, or

    (*c*) components for systems of heating, lighting, air-conditioning, ventilation, power supply, drainage, sanitation, water supply or fire protection, or for security or communications systems.

(4) In this Act references to construction operations do include a case where the things referred to in *subsection (3)* are supplied under a contract which also provides for their installation.

2.—(1) A contract is not a construction contract—

    (*a*) if the value of the contract is not more than €10,000, or

    (*b*) if—

        (i)   the contract relates only to a dwelling, and

        (ii) the dwelling has a floor area not greater than 200 square metres, and

        (iii) one of the parties to the contract is a person who occupies, or intends to occupy, the dwelling as his or her residence.

(2) A contract of employment (within the meaning of the Organisation of Working Time Act 1997) is not a construction contract.

(3) A contract between a State authority and its partner in a public private partnership arrangement, as those terms are defined in the State Authorities (Public Private Partnership Arrangements) Act 2002, is not a construction contract.

(4) Where a contract contains provisions in relation to activities other than those referred to in the definition of a construction contract and *section 1(2)*, it is a construction contract only so far as it relates to those activities.

(5) This Act applies to a construction contract whether or not—

    (*a*) the law of the State is otherwise the applicable law in relation to the construction contract, or

    (*b*) the parties to the construction contract purport to limit or exclude its application.

3.—(1) A construction contract shall provide for—

    (*a*) the amount of each interim payment to be made under the construction contract, and

    (*b*) the amount of the final payment to be made under the construction contract,

or for an adequate mechanism for determining those amounts.

(2) A construction contract shall provide for—

(a) the payment claim date, or an adequate mechanism for determining the payment claim date, for each amount due under the construction contract, and

(b) the period between the payment claim date for each such amount and the date on which the amount is so due.

(3) The *Schedule* shall apply to a main contract if and to the extent that it does not make provision for the matters specified in *subsections (1)* and *(2)*.

(4) The *Schedule* shall apply to a subcontract except to the extent that it makes provision which is more favourable to the executing party than that which would otherwise be made by the *Schedule*.

(5) Except after the occurrence of the circumstances specified in *subsection (6)*, a provision in a construction contract is ineffective to the extent that it provides that payment of an amount due under the construction contract, or the timing of such a payment, is conditional on the making of a payment by a person who is not a party to the construction contract.

(6) The circumstances referred to in *subsection (5)* are:

(a) where the other person is a company other than an unregistered company—

(i) the commencement of its winding up pursuant to section 251 of the Companies Act 1963 where no declaration of solvency has been made under section 256 of the Companies Act 1963,

(ii) the presentation of a petition to wind it up pursuant to section 213 of the Companies Act 1963,

(iii) the appointment of a receiver in respect of any of its property or assets, or

(iv) the presentation of a petition for the appointment of an examiner under the Companies (Amendment) Act 1990 in relation to it;

(b) where the other person is an unregistered company, the commencement of its winding up pursuant to section 345 of the Companies Act 1963;

(c) where the other person is an individual or partnership, the making of an application for adjudication under the Bankruptcy Act 1988 in relation to it;

(d) the making of a winding up or similar order by a court in relation to the other person;

(e) the occurrence of any event corresponding to those specified in this subsection under the law of any state to which Council Regulation (EC) No. 1346/2000 of 29 May 2000[1] on insolvency proceedings applies.

4.—(1) This section applies where, not later than 5 days after the payment claim date, an executing party to a construction contract delivers a payment claim notice relating to a payment claim to the other party or another person specified under the construction contract.

(2) A payment claim notice is a notice specifying—

(*a*) the amount claimed (even if the amount is zero),
(*b*) the period, stage of work or activity to which the payment claim relates,
(*c*) the subject matter of the payment claim, and
(*d*) the basis of the calculation of the amount claimed.

(3) If the other party or specified person referred to in *subsection (1)* contests that the amount is due and payable, then the other party or specified person—

(*a*) shall deliver a response to the payment claim notice to the executing party, not later than 21 days after the payment claim date, specifying—

 (i)   the amount proposed to be paid,
 (ii) the reason or reasons for the difference between the amount in the payment claim notice and the amount referred to in *subparagraph (i)*, and
 (iii) the basis on which the amount referred to in *subparagraph (i)* is calculated,

and

(*b*) if the matter has not been settled by the day on which the amount is due, shall pay the amount referred to in *paragraph (a)* to the executing party not later than on that day.

(4) Where a reason for the different amount in the response is attributable to a claim for loss or damage arising from an alleged breach of any contractual or other obligation of the executing party (under the construction contract or otherwise), or any other claim that the other person alleges against the executing party, the response shall also specify—

(*a*) when the loss was incurred or the damage occurred, or how the other claim arose,
(*b*) the particulars of the loss, damage or claim, and
(*c*) the portion of the difference that is attributable to each such particular.

(5) The rights and obligations conferred or imposed by this section are additional to any conferred or imposed by the terms of the construction contract.

5.—(1) Where any amount due under a construction contract is not paid in full by the day on which the amount is due, the executing party may suspend work under the construction contract by giving notice in writing under *subsection (2)*.

(2) Notice under this subsection shall specify the grounds on which it is intended to suspend work and shall be delivered to the other party—

(*a*) not earlier than the day after the day on which the amount concerned is due, and

(*b*) at least 7 days before the proposed suspension is to begin.

(3) Work may not be suspended under *subsection (1)*—

(*a*) after payment by the other party of the amount due, or

(*b*) after notice has been served by a party to the construction contract under *section 6(2)* in relation to a dispute relating to payment of the amount concerned.

(4) Where work is suspended under *subsection (1)* and the ability of the executing party to complete work within a contractual time limit is affected by the suspension of work, the period of suspension shall be disregarded for the purpose of computing the contractual time limit unless the suspension of work is unjustified in the circumstances.

(5) Where work is suspended under *subsection (1)* and the ability of a subcontractor to complete work within a contractual time limit is affected by the suspension of work, the period of suspension shall be disregarded for the purpose of computing the contractual time limit.

(6) A period of suspension of work under *subsection (1)* shall also be disregarded for the purpose of computing the time taken to complete the work under another construction contract where—

(*a*) the construction contract the work under which is suspended is a subcontract,

(*b*) the other construction contract is also a subcontract and the other party to that other subcontract is the same as the other party to the subcontract the work under which is suspended, and

(*c*) the ability of the executing party under that other subcontract to complete work within a contractual time limit is affected by the suspension of work.

(7) This section is without prejudice to the right of the other party to the construction contract under which work is suspended to claim for compensation or damages for any loss due to a suspension of work that is unjustified in the circumstances.

6.—(1) A party to a construction contract has the right to refer for adjudication in accordance with this section any dispute relating to payment arising under the construction contract (in this Act referred to as a "payment dispute").

(2) The party may exercise the right by serving on the other person who is party to the construction contract at any time notice of intention to refer the payment dispute for adjudication.

(3) The parties may, within 5 days beginning with the day on which notice under *subsection (2)* is served, agree to appoint an adjudicator of their own choice or from the panel appointed by the Minister under *section 8*.

(4) Failing agreement between the parties under *subsection (3)*, the adjudicator shall be appointed by the chair of the panel selected by the Minister under *section 8*.

(5) The party by whom the notice under *subsection (2)* was served—

(*a*) shall refer the payment dispute to the adjudicator within 7 days beginning with the day on which the appointment is made, and

(*b*) shall at the same time provide a copy of the referral and all accompanying documents to the person who is party to the construction contract.

(6) The adjudicator shall reach a decision within 28 days beginning with the day on which the referral is made or such longer period as is agreed by the parties after the payment dispute has been referred.

(7) The adjudicator may extend the period of 28 days by up to 14 days, with the consent of the party by whom the payment dispute was referred.

(8) The adjudicator shall act impartially in the conduct of the adjudication and shall comply with the code of practice published by the Minister under *section 9*, whether or not the adjudicator is a person who is a member of the panel selected by the Minister under *section 8*.

(9) The adjudicator may take the initiative in ascertaining the facts and the law in relation to the payment dispute and may deal at the same time with several payment disputes arising under the same construction contract or related construction contracts.

(10) The decision of the adjudicator shall be binding until the payment dispute is finally settled by the parties or a different decision is reached on the reference of the payment dispute to arbitration or in proceedings initiated in a court in relation to the adjudicator's decision.

(11) The decision of the adjudicator, if binding, shall be enforceable either by action or, by leave of the High Court, in the same manner as a judgment or order of that Court with the same effect and, where leave is given, judgment may be entered in the terms of the decision.

(12) The decision of the adjudicator, if binding, shall, unless otherwise agreed by the parties, be treated as binding on them for all purposes and may accordingly be relied on by any of them, by way of defence, set-off or otherwise, in any legal proceedings.

(13) The adjudicator may correct his or her decision so as to remove a clerical or typographical error arising by accident or omission but may not reconsider or re-open any aspect of the decision.

(14) The adjudicator is not liable for anything done or omitted in the discharge or purported discharge of his or her functions as adjudicator unless the act or omission is in bad faith, and any employee or agent of the adjudicator is similarly protected from liability.

(15) Each party shall bear his or her own legal and other costs incurred in connection with the adjudication.

(16) The parties shall pay the amount of the fees, costs and expenses of the adjudicator in accordance with the decision of the adjudicator.

(17) An adjudicator may resign at any time on giving notice in writing to the parties to the dispute and the parties shall be jointly and severally liable for the payment of the reasonable fees, costs and expenses incurred by the adjudicator up to the date of resignation.

(18) The parties to a dispute may at any time agree to revoke the appointment of the adjudicator and the parties shall be jointly and severally liable for the payment of the reasonable fees, costs and expenses incurred by the adjudicator up to the date of the revocation.

7.—(1) Where any amount due pursuant to the decision of the adjudicator is not paid in full before the end of the period of 7 days beginning with that on which the decision is made, the executing party may suspend work under the construction contract by giving notice in writing under *subsection (2)*.

(2) Notice under this subsection shall specify the grounds on which it is intended to suspend work and shall be delivered to the other party not later than 7 days before the proposed suspension is to begin.

(3) Work may not be suspended under *subsection (1)*—

    (*a*) after payment by the other party of the amount due, or
    (*b*) after the decision of the adjudicator is referred to arbitration or proceedings are otherwise initiated in relation to the decision.

(4) Where work is suspended under *subsection (1)* and the ability of the executing party or a subcontractor to complete work within a contractual time limit is affected by the suspension of work, the period of suspension shall be disregarded for the purpose of computing the contractual time limit.

(5) A period of suspension of work under *subsection (1)* shall also be disregarded for the purpose of computing the time taken to complete the work under another construction contract where—

(a) the construction contract, the work under which is suspended, is a subcontract,

(b) the other construction contract is also a subcontract and the other party to that other subcontract is the same as the other party to the subcontract the work under which is suspended, and

(c) the ability of the executing party under that other subcontract to complete work within a contractual time limit is affected by the suspension of work.

8.—(1) The Minister shall from time to time select persons to be members of a panel (in this section referred to as the "panel") to act as adjudicators in relation to payment disputes and shall select one of those persons to chair the panel.

(2) Persons selected under *subsection (1)* shall be members of the panel for a period of 5 years commencing on the date of selection and shall be eligible for reselection at the end of the period of 5 years.

(3) The Minister may, for good and sufficient reason, remove a member of the panel.

(4) A member of the panel may at any time resign by giving notice in writing to the Minister.

(5) In selecting persons to be members of the panel, the Minister shall have regard to their experience and expertise in dispute resolution procedures under construction contracts; and a person may not be selected to be a member of the panel unless the person is a person of any of the descriptions specified in *subsection (6)*.

(6) The descriptions of persons referred to in *subsection (5)* are as follows:

(a) a registered professional as defined in section 2 of the Building Control Act 2007;

(b) a chartered member of the Institution of Engineers of Ireland;

(c) a barrister;

(d) a solicitor;

(e) a fellow of the Chartered Institute of Arbitrators;

(f) a person with a qualification equivalent to any of those specified in *paragraphs (a)* to *(e)* duly obtained in any other Member State of the European Union.

9.—The Minister may prepare and publish a code of practice governing the conduct of adjudications under *section 6*.

10.—(1) The parties to a construction contract may agree on the manner by which notices under this Act shall be delivered.

(2) If or to the extent that there is no such agreement, a notice may be delivered by post or by any other effective means.

(3) Where under this Act a notice is required to be delivered not later than a specified number of days after a particular date and the last of those days is a day which is a Saturday or Sunday or a public holiday (within the meaning of the Organisation of Working Time Act 1997), the notice shall be taken to be validly delivered if delivered on the next day which is not such a day.

11.—The expenses incurred by the Minister in the administration of this Act shall be paid out of moneys provided by the Oireachtas.

12.—(1) This Act may be cited as the Construction Contracts Act 2013.

(2) This Act applies in relation to construction contracts entered into after such day as the Minister may by order appoint.

SCHEDULE   *Section 3.*

Provisions to Apply to Matters Regarding Payments

1. The payment claim dates under a construction contract shall (subject to *paragraph 2*) be as follows:

   (*a*) 30 days after the commencement date of the construction contract;
   (*b*) 30 days after the date referred to in *clause (a)* and every 30 days thereafter up to the date of substantial completion;
   (*c*) 30 days after the date of final completion.

2. Where a construction contract provides, or the parties to a con-struction contract otherwise agree, that the duration of the work under the construction contract is or is estimated to be less than 45 consecutive days, the payment claim date shall be 14 days following completion of the work under the construction contract.

3. The date on which payment is due in relation to an amount claimed under a construction contract shall be no later than 30 days after the payment claim date.

4. The amount of an interim payment under a construction contract shall (subject to *paragraph 5*) be the difference between—

   (*a*) the aggregate of the gross value (determined in accordance with the con-struction contract) of the work done under the construction contract at the payment claim date concerned together with any additional amounts in the interim payment under the construction contract, less any deductions from payment provided for by the construction contract, and
   (*b*) the aggregate amount of interim payments that have already been made at that payment claim date.

5. The aggregate of payments made under a construction contract shall not exceed—

- (*a*) the amount provided for in the construction contract as originally concluded, and
- (*b*) amounts provided for by any amendments to that contract agreed between the parties.

Reference

1   OJ No. L160, 30 June 2000, p.1

Appendix B

Code of Practice Governing the Conduct of Adjudications under Section 6 of the Construction Contracts Act 2013

This Code of Practice is made pursuant to section 9 of the Construction Contracts Act, 2013 and should be read in conjunction with that Act.

Definitions

1. A reference in this Code of Practice to:

    a) "the Act" means the Construction Contracts Act, 2013;

    b) "Adjudicator" means an Adjudicator who is appointed to a payment dispute in accordance with section 6 of the Act;

    c) "Chairperson" means the Chairperson of the Construction Contracts Adjudication Panel who is appointed by the Minister under section 8(1) of the Act;

    d) "Construction Contracts Adjudication Service" means the section of the Department of Jobs, Enterprise and Innovation responsible for, inter alia, processing applications to the Chairperson under section 6(4) of the Act and contact details are available at www.djei.ie;

    e) "Minister" means the Minister or Minister of State with responsibility for the Construction Contracts Act, 2013;

    f) "Notice of Intention" means the notice of intention to refer a payment dispute for adjudication, referred to in section 6(2) of the Act;

    g) "Panel" means the panel of Adjudicators referred to in section 8(1) of the Act, the members of which are appointed by the Minister; and

    h) "Payment dispute" has the meaning assigned to it by section 6 of the Act.

General

2. The procedures set out in this Code of Practice shall apply to each individual payment dispute arising under the Act. In accordance with section 6(9) of the Act, an Adjudicator may deal at the same time with several payment disputes arising under the same construction contract or related construction contracts.

3. No liability whatsoever shall extend to the Minister, Chairperson or to the Department of Jobs, Enterprise and Innovation in respect of this Code of

Practice or for any loss that arises from the operation of this Code of Practice. The Minister reserves the right to make changes to this Code of Practice.

Preliminary

4.   A party to the construction contract (known as "the Referring Party") commences adjudication pursuant to section 6(2) of the Act by serving a written Notice of Intention on the other party or parties to the construction contract (known as the "Responding Party/Parties") under which an individual payment dispute arises.

5.   A Notice of Intention shall include:

   (i)   the name, address and contact details of each party to the construction contract;
   (ii) relevant details of the payment dispute to include the amount in dispute (even if the amount is zero), the nature of the payment dispute, and the site address;
   (iii) a copy of the relevant payment claim notice, and any response to that payment claim notice as provided for in section 4 of the Act; and
   (iv) relevant details to identify the construction contract and any supporting information that may assist an Adjudicator in understanding the nature of the payment dispute. Where a written construction contract exists, this must be attached.

Prospective Adjudicator responsibilities to the parties to a payment dispute

6.   A prospective Adjudicator should only accept an appointment to a payment dispute under the Act if he/she:

   (i)   is able to give the adjudication the time and attention which the parties to the payment dispute are reasonably entitled to expect;
   (ii) believes that he/she is competent to determine the issues in dispute; and
   (iii) is satisfied that no conflict of interest exists between him/her and the parties subject to paragraphs 11 and 20 of this Code of Practice.

7.   A prospective Adjudicator shall not contact any party to a payment dispute under the Act in order to solicit appointment as an Adjudicator to that dispute.

The Appointment of an Adjudicator – by agreement of the parties

8.   The parties to the construction contract may, within five days beginning with the day on which Notice of Intention is served, agree to appoint an Adjudicator of their own choice and he/she may be a person referred to in the construction contract to perform that role, a person from the Panel or he/she may be another suitably qualified person.

9. A person who is requested to accept an appointment as Adjudicator following an agreement by the parties to the construction contract in accordance with section 6(3) of the Act shall, within two days of such a request and prior to accepting the appointment, write to the parties to ask them to disclose any information indicating any potential conflict of interest that may arise from the person's appointment as Adjudicator. He/she shall draw the attention of the parties to the provisions of paragraph 32 of this Code of Practice. The prospective Adjudicator shall, at the same time provide the parties with his/her proposed terms and conditions of appointment, including the basis for his/her fees, costs and expenses.

10. Each party shall within three days of the communication from the prospective Adjudicator decide if the appointment of the prospective Adjudicator is to proceed and inform the prospective Adjudicator in writing of their decision.

11. If a potential conflict of interest is disclosed by any party, the prospective Adjudicator may subject to the consent of all the parties, and on satisfying any professional and/or ethical concerns he/she may have, accept the appointment.

12. If the appointment of the prospective Adjudicator is to proceed, the prospective Adjudicator shall write to each party to accept the appointment and the date of the letter of acceptance sent to the parties shall be deemed to be the date on which the appointment of the Adjudicator is made for the purposes of section 6(5)(a) of the Act. Such acceptance, anonymised in terms of the details of the parties to the dispute, shall be notified by the Adjudicator to the Construction Contracts Adjudication Service of the Department of Jobs, Enterprise and Innovation for the purpose of compiling statistical information relating to the Act.

The Appointment of an Adjudicator – by the Chairperson

13. Failing agreement by the parties to select an Adjudicator in accordance with section 6(3) of the Act, a party to the construction contract may apply to the Chairperson to seek the appointment of an Adjudicator from the Panel in accordance with section 6(4) of the Act. Relevant contact details are available on the website of the Department of Jobs, Enterprise and Innovation at www.djei.ie.

14. If an application is to be made under section 6(4) of the Act to the Chairperson, it shall be made not earlier than five days from and including the day on which the Notice of Intention was served.

15. An application to the Chairperson to appoint an Adjudicator from the Panel to a payment dispute shall be in writing and submitted to the Chairperson in accordance with the application procedures set out by the Construction Contracts Adjudication Service of the Department of Jobs Enterprise and Innovation from time to time. Such application, shall be copied by the

applicant to the other party/parties to the payment dispute at the same time and shall include:

(i)   the name, address and contact details of each party to the construction contract;

(ii) relevant details of the payment dispute to include the amount in dispute (even if the amount is zero), the nature of the payment dispute, and the site address;

(iii) a copy of the Notice of Intention including any accompanying documents attached to that Notice;

(iv) the date as to when the Notice of Intention was served on the Responding Party/Parties and how this was done; and

(v)   relevant details to identify the construction contract and any supporting information that may assist an Adjudicator in understanding the nature of the payment dispute. Where a written construction contract exists, this must be attached.

16. The Chairperson and/or the Construction Contracts Adjudication Service of the Department of Jobs, Enterprise and Innovation may seek further information or clarification(s) from the applicant relevant to the nature of the dispute and such information or clarification(s) should be provided promptly by the applicant and copied to the other party/parties to the payment dispute at the same time. No additional or other supporting information should be submitted by the applicant without a specific request for such information from the above-mentioned in this paragraph.

17. The Chairperson shall, following receipt of a completed application from a party to the construction contract made in accordance with paragraph 15 of this Code of Practice and subject to paragraph 16 of this Code of Practice, appoint an Adjudicator from the Panel.

18. The appointment of an Adjudicator from the Panel shall be made by the Chairperson and notified in writing by the Construction Contracts Adjudication Service of the Department of Jobs, Enterprise and Innovation to the parties, normally within seven days after the receipt of the application to the Chairperson, subject to paragraph 16 of this Code of Practice. The date of the letter from the Construction Contracts Adjudication Service to the parties shall be deemed to be the date on which the appointment of the Adjudicator is made for the purposes of section 6(5) (a) of the Act.

19. An Adjudicator appointed by the Chairperson to a payment dispute shall, within two days of such appointment, request of the parties in writing to disclose any information indicating any potential conflict of interest that may arise from the person's appointment as Adjudicator. He/she shall draw the attention of the parties to the provisions of paragraph 32 of this Code of Practice. The Adjudicator shall at the same time provide the parties with his/her terms and conditions of appointment, including the basis for his/her fees, costs and expenses.

20. If the information disclosed indicates a potential conflict of interest, the Adjudicator may only proceed with the adjudication where he/she is satisfied that the disclosures are frivolous or vexatious; that no professional or ethical concerns arise; and that no actual conflict of interest exists.

Referral of a payment dispute to an Adjudicator

21. Following the appointment of an Adjudicator, the Referring Party shall in accordance with section 6(5) of the Act refer the payment dispute to the Adjudicator in writing within seven days of the Adjudicator's appointment, and the Referring Party shall provide a copy of all such documentation to the Responding Party/Parties at the same time.

22. The referral of the payment dispute to the Adjudicator shall include:

(i) the name, address and contact details of each party to the construction contract;

(ii) relevant details of the payment dispute to include the amount in dispute (even if the amount is zero), the nature of the payment dispute, and the site address;

(iii) a copy of the Notice of Intention including any accompanying documents attached to that Notice;

(iv) the date when the Notice of Intention was served on the Responding Party/Parties and how this was done;

(v) the contentions on which the Referring Party intends to rely upon to support their case; and

(vi) relevant details to identify the construction contract and any supporting information that may assist an Adjudicator in understanding the nature of the payment dispute. Where a written construction contract exists, this must be attached.

Adjudication of a payment dispute – Procedures and Decision

23. The Adjudicator in any payment dispute under the Act shall be impartial, independent and only adjudicate where satisfied that no actual conflict of interest exists. He/she shall observe the principles of procedural fairness, which shall include giving each party a reasonable opportunity to put their case and to respond to the other party's case.

24. For the purposes of the adjudication proceedings, the Adjudicator may:

(i) request any reasonable supporting or supplementing documents pertaining to the payment dispute detailed in the Notice of Intention and/ or in the referral of the payment dispute to the Adjudicator;

(ii) take the initiative in ascertaining the facts and matters required for a decision and make use of his/her own specialist knowledge, if it is appropriate to do so. If the Adjudicator uses any such specialist knowledge he/ she shall disclose this to the parties as appropriate;

(iii)   appoint experts, assessors or legal advisers, provided that the parties have been notified of their identity and their terms of reference;

(iv)   make site visits and inspections or carry out tests, subject to prior notification to the parties and obtaining any necessary consent from a third party or parties;

(v)   invite written submissions/representations and evidence from the parties, if appropriate;

(vi)   meet jointly with the parties and their representatives, if any, to enable further investigation;

(vii)   hold a teleconference with the parties, with the consent of the parties; and

(viii)   hold an oral hearing, where appropriate.

25. The Adjudicator shall upon receipt of the referral of a payment dispute from the Referring Party, inform the parties in writing of the date on which it was received by the Adjudicator. This date of receipt shall be regarded as the date on which the referral of the payment dispute to the Adjudicator has been made for the purposes of section 6(6) of the Act.

26. The Adjudicator shall, at the same time, also inform the parties of the procedures that he/she intends to apply during the adjudication process. This shall include directions as to the timetable for the adjudication and any deadlines to be adhered to by the parties and/or limits as to the length of written documents. The Adjudicator shall draw the attention of the parties to the provisions of paragraph 32 of this Code of Practice. The Adjudicator may revise his/her guidance to the parties on the above mentioned matters in circumstances where he/she considers it necessary to do so and he/she shall inform the parties of any such change as appropriate.

27. The Adjudicator shall ensure that the procedure adopted is commensurate with the nature and value of the payment dispute and he/she shall be mindful of whether or not an oral hearing is required having regard to matters such as to whether or not there is a conflict of fact or other relevant matter that requires such a hearing.

28. The Adjudicator shall use reasonable endeavours to process the payment dispute between the parties in the shortest time and at the lowest cost. He/she shall promptly notify the parties of any matter that will slow down or increase the cost of making a determination.

29. The parties may agree to revoke the appointment of the Adjudicator in accordance with section 6(18) of the Act and shall be jointly and severally liable for the payment of the reasonable fees, costs and expenses incurred by the Adjudicator up to the date of the revocation.

30. In accordance with section 6(17) of the Act, the Adjudicator may resign for reasonable cause at any time on giving notice in writing to the parties to the payment dispute. Such resignation, anonymised in terms of the details of the parties to the dispute, shall be notified by the Adjudicator

to the Construction Contracts Adjudication Service of the Department of Jobs, Enterprise and Innovation for the purpose of compiling statistical information relating to the Act.

31. Upon such resignation the adjudication is at an end subject to the payment by the parties of the reasonable fees, costs and expenses incurred by the Adjudicator up to the date of resignation. The parties shall be jointly and severally liable for the payment of the reasonable fees, costs and expenses incurred by the Adjudicator up to the date of resignation.

32. If a party to the adjudication, without showing sufficient cause, fails to:

    (i)   attend a meeting; or
    (ii)   comply with any directions of the Adjudicator made in accordance with paragraph 26 of this Code of Practice; or
    (iii) disclose any information indicating a potential conflict of interest as required to do in accordance with paragraphs 9 and 19 of this Code of Practice; or
    (iv)   produce any document or written statement requested by the Adjudicator;

    the Adjudicator may:

    a)   continue the adjudication in the absence of a party;
    b)   continue the adjudication without the document or written statement requested;
    c)   draw such inferences from that failure to comply as circumstances may, in the Adjudicator's opinion, be justified;
    d)   make a decision on the basis of the material properly provided; and
    e)   make a decision apportioning the fees, costs and expenses of the Adjudicator, as appropriate.

33. The Adjudicator shall, in accordance with the Act, reach a decision within 28 days beginning with the day on which the referral is made or such longer period as is agreed by the parties after the payment dispute has been referred. The Adjudicator may extend the period of 28 days by up to 14 days, with the consent of the Referring Party.

34. The decision of the Adjudicator on a payment dispute shall be in writing and it shall be signed and dated by the Adjudicator. Unless the parties agree otherwise in writing, the decision shall include the reasons for the decision.

35. The Adjudicator's decision shall allocate such fees, costs and expenses of the Adjudicator as he/she has authority to allocate under section 6(16) of the Act and under the provisions of this Code of Practice.

36. The Adjudicator's fees, costs and expenses shall be reasonable in amount having regard to the amount in dispute, the complexity of the dispute, the time spent by the Adjudicator and other relevant circumstances.

37. Any document or information supplied for and/or disclosed in the course of the adjudication shall be kept confidential by the Adjudicator. He/she will only disclose such document or information if required to do so by law, or pursuant to an order of a court, or with the consent of all the parties to the payment dispute.

38. The parties are responsible for their own legal and other costs incurred in connection with the adjudication in accordance with section 6(15) of the Act.

Reporting on the Conduct of Adjudication Cases

39. The Chairperson may seek or, put in place arrangements to seek, details of adjudication cases from Adjudicators and which shall not include the names of the parties to a payment dispute. An Adjudicator, regardless of whether appointed to a payment dispute under section 6(3) or 6(4) of the Act, shall provide such anonymised information to the Construction Contracts Adjudication Service of the Department of Jobs, Enterprise and Innovation on each adjudication case within 21 days of the completion of the case. This will be used for the purpose of compiling statistical information relevant to adjudications conducted in accordance with the Act.

I, PAT BREEN, Minister of State at the Department of Jobs, Enterprise and Innovation, in exercise of the powers conferred on me by section 9 of the Construction Contracts Act, 2013 (No. 34 of 2013) and the Construction Contracts (Transfer of Departmental Administration and Ministerial Functions) Order 2014 (S.I. No. 476 of 2014) (amended by the Construction Contracts (Transfer of Departmental Administration and Ministerial Functions) Order 2015 (S.I. No. 173 of 2015)) and the Jobs, Enterprise and Innovation (Delegation of Ministerial Functions) Order 2016 (S.I. No. 333 of 2016) make this Code of Practice.

The Construction Contracts Act 2013 Code of Practice Governing the Conduct of Adjudications made on the 5th day of July 2016 is revoked.

PAT BREEN,
Minister of State at the Department of Jobs, Enterprise and Innovation.

Date: 25 July 2016

Index

For details of the legal cases referred to in the text please see separate *Table of cases*.